Fabio Tabbo'

Stop–Smoking
Mentalism

Close-up Hypnosis
Vol. 1

Published by Underwords
207 - 1425 Marine Drive
West Vancouver, BC – Canada V7T 1B9
Tel. (toll-free line): +1-866-308-3388
Fax: +1-604-677-7476
Email: info@underwords.com
Website: http://www.underwords.com

1st Edition
ISBN 0-9739456-0-5
ISSN 1715-8486 (Close-up Hypnosis Vol. 1)

To my family,
pale blue eyes,
vomiting dolls,
lost friends.

TABLE OF CONTENTS

ACKNOWLEDGMENTS

Special thanks to the following contributors:
(in alphabetical order)

Dr. Nicola Dexter
(Session trance-script)
Health & Happiness
37 Hartland Road
Camden, London - UK NW1 8DB
Tel. +44-207-2674706 • mob. +44-7711-462923
fax +44-207-8133769
Email: hypno@nicoladexter.com
Website: http://www.nicoladexter.com

Dr. Maurice Kouguell
(Scripts adjustments)
Brookside Center for Counseling and Hypnotherapy
20 Chestnut Street, Suite 15N
Exeter, New Hampshire 03833 - USA
Tel/fax +1-603-7789955
Email: hypnoprof@comcast.net
Website: http://www.brooksidecenter.com

Mark Tyrrell
(General induction)
Uncommon Knowledge Ltd.
12 Queen Square
Brighton - UK BN1 3FD
Tel. +44-01273-776770 • fax +44-01273-774427
Website: http://www.uncommon-knowledge.co.uk

Additional comments or suggestions:
**Bruce Bernstein, Chuck Hickok, Elizabeth Loftus,
Bill Palmer, George Robinson Jr.**

Inside pictures: **M-Cat Studio**

Hand model: **Floriane Margheret**

Cover design: **Rebecca Raupp**

Cover photograph: **Linda R.D. Robinson**

Model: **Norman Robinson**

Typesetting: **Robin Cummings**

No-smoking campaign pictures ©
Audiovisual Library of the European Commission
http://www.eu.int

Additional illustration ©
CORNE (Santiago Cornejo)
http://www.corne.com.ar

"Smoking Cessation" audio file ©
Jef Gazley, LMFT
http://www.asktheinternettherapist.com and
http://www.hypnosistapes4health.com

*Words are the most
powerful drug used by mankind.*

Rudyard Kipling

PREFACE

T hank you for taking the time to peek into my little warped world.

Even if you don't frequent circles where guns are more tolerable than cigarettes, smoking is always a controversial topic, with immediate appeal on any type of audience. Among the large variety of treatments available, hypnosis is certainly one of the most fascinating and widely misunderstood: this book will allow you to present an impromptu demonstration of no-smoking hypnosis, in informal situations or as part of a formal session, both in close-up and in parlor/stage settings.

The methods and scripts you are about to read were initially devised for my own performances. For the sake of completeness, I later added the detailed descriptions of the patters, movements and silences, and explained why I had ultimately changed a specific keyword in the script, or gestured in a particular way during the effect, or increased the pace of the routine at a given moment. I also tried to acknowledge in the most accurate way the authors who had influenced me – the hardest part of the artistic creation is to duly credit your sources of inspiration, or to dishonestly conceal them...

Finally, "Rewriting". Robert Louis Stevenson once stated, "When I say *writing*, believe me, it is *rewriting* I have chiefly in mind". And in our mental world, with all due proportions, an effect is never truly finalized: it is only abandoned for performance or publication purposes. The volume you are now holding is the result of over ten years of constant improvements and evolving doubts... I wish you the best of luck with it.

Fabio

Accra, October 2005

INTRODUCTION: IN EFFECT

spectator is asked to participate in an experiment of no-smoking hypnosis. She pretends to smoke an imaginary cigarette, and feels as if a real one was being used: she vividly visualizes its gray smoke, smells and tastes the tobacco, feels like coughing etc. She then brings this cigarette closer to her hand, causing a noticeable warming sensation; and when she finally decides to stub it out on her hand, she feels a sharp burn!

The effect is achieved through the use of secret gimmicks, hypnotic language patterns, psychological subtleties and stealthy physical manipulations; however, it appears impromptu, propless and gimmickless to the audience, with the performer away at all times.

The main routine will be described in minute details, while the variations will only highlight the differences with the original version. The explanation of the theoretical principles won't be separated from the performance analysis, but seamlessly integrated within the description of each relevant phase. Before examining the gimmicks and the actual effect, let us describe the general guidelines that will be followed in this first effect and in all forthcoming volumes.

I. GENERAL PRINCIPLES

The *Close-up Hypnosis* Collection, as the name suggests, aims at showcasing a number of mental performance pieces with a hypnotic theme, achievable in real world impromptu situations. All routines will follow the set of guidelines summarized below.

I.1 Mentalism and hypnosis

The effects focus on gimmicked hypnosis and "real" suggestive mentalism, but could also be presented as pseudo-psychic feats or extra-sensory exploits. They can be played seriously as real hypnotic experiments, or inserted into a mentalism show, both in close-up and in formal stage settings. The blend of eclectic techniques drawn from the fields of mentalism, hypnosis, magic, psychology, linguistics and Neuro-Linguistic Programming (NLP) allows the performer to reframe the events in many different ways.

Certain "mental magic" effects have such a strong presentational appeal and entertainment value that they will find their way through the Collection – as long as they possess that "overwhelming hypnotic theme", to paraphrase Richard Busch in *Peek Performances*. What you will NOT find in these pages are compromises belonging to the magic tricks' realm, such as "The spectator extracts a paper bag from a mahogany box and selects a poker chip used to determine the page of a magazine, which is torn into pieces until one freely chosen word matches the one written on an initialed index card held in full view by a bulldog clip since the beginning of the show". Some bizarre presentations also possess heavy-hitting features and will be showcased in the series. I initially planned to leave more space to the "weerd"[1] side, but I was told too many dreadful jokes (e.g. a baby vampire bit a sandwitch, he was coffin' and scrying for his mummy etc.) and needed to take some distance for a while.

I.2   Seemingly impromptu

As opposed to "simply" impromptu. The effects may often require some secret gimmicks (usually prepared prior to performance, or even before leaving your home), hence the qualifier "seemingly": the effects will *seem*, from the audience point of view, performed at a moment's notice and free of any staged trickery. They must look totally unprepared and off the cuff, even if you spent days in setting up your gimmicks and have been carrying them around pointlessly, waiting for the right performance opportunity. The struggles you had in building this servante or locating that rare chemical simply don't matter - as long as the audience will never be aware of it. Winston Churchill once said that people just wanted us to bring the gold home: they didn't care about the troubles we encountered during the travel, or the pains we endured to achieve our goals.

---

[1] Special thanks to Docc Hilford for the misspelling.

Of course the effect could really be impromptu, without any setup or gimmick, but the utilization of a secret gizmo generally allows you to be one step ahead and present a stronger effect, reducing the "lightweight" factor (see rule #10). As per the Joseph Dunninger's line of thought, if props lower your price, hidden gimmicks increase it.

I.3   Propless

Not necessarily gimmickless, but propless. I do not feel guilty about stealthy accessories and secret setups, but strive to present an effect without any incongruent item. The introduction of as few external elements as possible lets the audience direct all their attention on what they think is the main component of the effect: it will be easier for them to follow the routine and apparently "control" the performer's moves, while in fact focused spectators are easier to manage – hence to misdirect.

Since we have to work with the audience in mind, an effect can be considered "propless" when the spectators are not aware of any prop (in the same way as a routine is considered impromptu when no preparation is perceived by the audience). I consider a "prop" any incongruent object <u>from a layman's perspective</u>: for example, anything that a spectator would not carry on her person in a specific performance environment. One does not necessarily need to look at large flashy boxes or Plexiglas stands: an envelope and a marker are indeed props, unless performing in an office or home where those objects are readily available. Envelopes and markers are certainly common household items, but how many laymen carry them while dining in a restaurant or walking down the street? In an informal setting, such as your friend's living room, bringing out those props could be perceived as incongruent and create an instinctive impression of trickery. And when the audience realizes that you are carrying around props for *planned,* tricky purposes, the effect cannot be called "impromptu" anymore; it is impromptu for the self-delusional performer, but not for the laymen. Compare it with borrowing a business card and a pen instead.

Similarly, some innocent items can become props in the wrong hands: the above-mentioned business card would probably seem out of place if retrieved by a 7-year-old child from his classy leather wallet. A pack of cigarettes borrowed from a spectator is fine; one you take out of your own pocket is acceptable if you are a smoker, but it becomes a prop if the audience knows you don't smoke...

To assess how a prop will be perceived by the audience, we should then consider:

- the way the item is embedded into the routine;
- the reasons provided for its presence;
- its congruity in a specific performance setting;
- its consistency with the performer's persona;
- the alternatives available in the mind of the audience.

In any case, the performer will not bring out a wooden case containing 13 laminated hypnotic symbols inserted into clear plastic holders etc.; the performer brings out nothing, and borrows a common, congruent, examinable object (or he uses his own in the worst-case scenario). The audience will identify with it immediately, without any possible suspicion. Besides, the possibility of returning the borrowed object in its magically "frozen" state at the end of the show, as Michael Ammar repeatedly pointed out, enhances the overall impact of the effect and leaves the spectator with an unforgettable souvenir. You may retort that a mysterious-looking jewel introduced by the performer would attract more attention than a simple borrowed coin, and would further underline the difference with previous propless effects. You would be right, and those variations will also be considered when applicable.

Somebody is probably wondering about playing cards... Their place in mentalism is a widely discussed issue: entertainment for some, sacrilege for others. While their presence in hypnosis seems even more arguable, presentational variations and additional ideas using cards may be briefly described for the sake of completeness.

I.4   Close-up compliance

All routines can be presented in *real* close-up settings. This "real-life" rule addresses three limitations commonly found in self-defined close-up effects, namely:

- The distance requirements and restrictive angles that the performer often needs to respect.
- The unavoidable - and sometimes unpleasant - human interaction elements found in informal situations frequent interruptions, examination requests, comments of people speaking to each other etc.
- The staged conditions occasionally required to set up the effects.

Of course, as a performance philosophy, remember the old saying: if you manage to make one person, only one person, wonder in awe during your show... then you're a pretty lousy entertainer. While the priority is given to intimate settings, most of the routines described in the Collection can also be performed for larger crowds in stand-up, parlor or even stage/TV conditions.

I.5   Real audience

You certainly realize that "real close-up" implies real people: breathing and breeding subjects who will interrupt you, contradict themselves and react in the most unpredictable ways. In my opinion, the Murphy's Law of Performance would sound like this: "In any group of spectators, of any age, any race, any religion, any socio-economical extraction, any political party, any sexual orientation and any cultural background, there will be *at least* one heckler".

To be on the safe side, the effects are devised for intelligent adults: demanding and suspicious people who will not allow you to stay alone in their living room before the show to psychically attune yourself to the surroundings, and who will not fall into deep trance during the first words of your induction. If by chance you find yourself in front of a crowd of gullible placators, excellent... you will be able to treat them with the same respect and consideration.

I.6   One-man show

"One-man show" means no performing partners, no assistants, no stooges - not even instant stooges. Ambiguous double-talks and uncontrolled dual-reality patters can lead to dangerous leakages of reputation in real close-up situations: people speak to each other, as opposed to stage hypnosis conditions... In other words, you are alone: before, during and after. To be honest, I occasionally enjoy some dual-reality drops here and there – but instilled in a way to minimize or emphasize specific aspects of the performance for a particular spectator (or group of spectators), without altering the overall perception of the effect.

I.7 No dress code

You probably don't fancy wearing unusual clothes or accessories that do not fit your usual dressing style (such as suspenders, or perhaps baggy trousers), just because you may be asked to perform a particular effect later in the day. Incongruent clothes can become props, and may cause suspicion in the eyes of spectators who know you well enough to notice the differences, unless you leave them enough time to be accustomed to your new garments. On the other hand, the "Can be performed naked" statement (to use a hype found in magic catalogs) is a bit extreme: apart from the obvious puritanical considerations, a pocket is often useful, as long as you keep it free from bulky props rattling and shaking inside.

Concerning the short-sleeves issue, which is more a magicians' concern, rolled-up sleeves usually suggest an air of openness; when performing mental magic (not "hypnosis") and when the circumstances allow it, I casually roll them up at the beginning of the show, in a "let's get to work" fashion.

I.8 Technical accessibility

You will need to perform some covert moves soon or late... You expected that, didn't you? Fortunately the effects do not rely on complicated manipulations or dexterity feats, but rather on secret gimmicks, hypnotic language patterns and psychological deceptions. This lets us focus on what really matters.

I.9 Tested and true

The Collection will only showcase practical, tested routines performed countless times in the real-life conditions described above. No theoretical hypnosis exaggeratedly simplifying the spectators' behavior, no abstract concepts for daydreaming magicians, no speculative principles applied to exemplificative tricks, no rhetorical conjuring for conjurers, no self-delusional pipedreams for armchair mentalists, no dogmatic rules from pretentious newcomers (...)

I.10 The lightweight factor

This is the most important rule. Let us for example consider a classic billet peek: this impromptu trick meets all the previous criteria, but the "impromptu=lightweight" equation may enter in (depending on the type of peek...). You know how it goes: you are asked to demonstrate your fabulous mental skills, right here, right now. You are caught unprepared but can still hear the voice of your mentor, reminding you that the simplest effects carry the strongest impact. And you have read that this new and improved peek wallet was a devastating killer, that people will NEVER suspect a peek after your revelation of the written information - they won't even notice the wallet. In fact, they won't even recall that you EVER touched the billet, and will eventually forget that anything was EVER written down! And at the end of the day, they won't remember ANYTHING AT ALL!! My friend, you are GOOD.

II. METHODOLOGICAL TOOLS

I n addition to the general Collection guidelines, let us briefly describe some methodological principles we will be using throughout the routines, along with my personal intentions and performance philosophy.

II.1 Assumptions

When creating an effect, I always consider the spectators' assumptions, the solutions they will devise to reverse-engineer its methodology and the ways of contradicting or reinforcing their conjectures. It is crucial to think like laymen, to walk in their shoes and relearn what they already know, what they suspect and what they expect from us.

The "Audience Assumptions" sections provide an overview of the laymen's standpoint, and include some of the queries and topics raised by the spectators after the show. People may have a genuine interest in the topics brought up by your effects, or they may just want to test your knowledge of the matter. For example, you just performed your favorite drawing duplication using a remote viewing presentation: impression pads or nail writers should be obscure subjects to the public, but they may link the effect to popular remote viewing stunts, and ask further details about that "famous" psychic detective who collaborated with the police and drew a sketch of the place where a kidnapper was holding her[2] prisoner. Failing to answer accordingly would not enhance your reputation. These sections will therefore provide a brief technical background on the supposed real work, and on the relevant issues raised by the audience.

As a side note, this idea came up after a rendition of a pendulum trick by Richard Webster: a participant began to ask me all sorts of questions about the momentum of the mass... At that time, I did not have the faintest idea of what she was blathering about, and thought that a preliminary research on the subject would have been time well spent. Of course I do not suggest you present yourself as a psychic with supernatural powers, or as a professional hypnotherapist (unless you are a certified practitioner) - it's just a matter of being informed, or at least pretending to be informed.

II.2   Audience-centered approach

I try as much as possible to create a positive personal experience customizable to most spectators. Your primary intention is probably entertainment (mystifying entertainment, that is), but why miss the chance of seeding positive suggestions? Hypnosis is certainly one fertile ground... "Entertain, Enlighten, Empower", said Jeff Mc Bride.

Since many of my effects are based on laymen's personal desires or needs (latent or specifically expressed), such as

---

[2] "Her prisoner" is simply a way of contradicting an existing assumption. It should also make you think differently.

stopping smoking or solving a personal issue, the positive suggestions come naturally embedded into the patter. Of course this approach should not be followed in all routines, as it would lead to a homogeneous, predictable and ultimately boring presentational pattern.

II.3 VAK

The VAK representational systems, as defined by NLP co-founders Richard Bandler and John Grinder, stand for the modes by which we represent external data – the three main ways of processing information: **V**isual (pictures, sights and images), **A**uditory (sounds, noises, tones and volumes) and **K**inesthetic (touch and pressure).

When we think about something, our thoughts are encoded by our senses, which make them more specific and "real". It is crucial to involve as many senses as possible during our performance, and maximize the spectators' participation by making the action happen in their hands (or in their minds). Furthermore, people like to touch and examine every single bit of what is used by the performer: let them do so whenever possible.

II.4 Devil in the details

Dariel Fitzkee in *Magic by Misdirection* stressed that illusions do not work without attention to detail. You may be familiar with ads stating that a book was including every single detail of a routine, that absolutely nothing had been left out, that the author had given away the whole store and the land it was sitting on, leaving no stone unturned etc. As mentioned in the introduction, and at the risk of sounding like a dull school teacher, I tried to describe every possible aspect of the performance, analyzing every keyword, every tiny movement, every pause and - more importantly - explaining their raison d'être. This will hopefully allow you to perform the effects in a proficient way, and save valuable time you would have required otherwise to integrate the scripts with additional information. Besides, a large number of subtleties described throughout the Collection are transferable to your existing repertoire: even if you don't have the time to learn new effects, you should be able to enhance the impact of your current routines. Of course the patters should be tailored to your own specifications, with the minor adjustments needed to fit your personality and your specific target audience: you are provided with the tools and the blueprints to adapt the scripts in an almost effortless way.

This attention to detail is justified by the following conviction: the way a sentence is phrased, or a question is asked, can make the difference between a "Yes" and a "No". And a persuaded, resounding "YES" from an audience member can suggest, presuppose, imply or validate a large variety of sensations, facts and memories. In her mind, or at least in the minds of the other spectators.

Concerning the visual details, I tried to insert as many illustrations as possible[3], to better describe the moves and accelerate the learning curve. A note of warning, though: even if all the performance details are plainly laid out, you may miss

---

[3] I even considered including the type of photographs shot by Tony "Doc" Shiels in *The Cantrip Codex*, to broaden the views and the market.

part of the contents... Reading between the lines is essential – small details require extra attention, and we only see what we deserve.

Certainly, deconstructing a hypnosis effect and analyze its working in such a detailed way may seem like dissecting a beetle: very few people will appreciate (the beetle itself might not find major benefits in it) and the rest will be bored to death. Hopefully the minimal marketing and the limited distribution of this book have only left the readers with a sincere interest in our Art(s).

II.5   Economy of language, economy of motion

Performers should consider both the economy of motion (minimizing movements during the effects - "What doesn't add, detracts") and the economy of language. To quote André Martinet's cynical approach on the matter (*Elements of General Linguistics*): "The economy of language is this permanent search for equilibrium between the contradictory needs which it must satisfy: communicative needs on the one hand and articulatory and mental inertia on the other".

Most people do not have the time to listen to you preaching for hours, unless you are a recognized hypnotherapist. I consequently cut and streamlined the scripts as much as possible, removing all unnecessary words and making them suitable for modern MTV-paced audiences whose primary concern is "What the **** do you have for me, NOW?!"

II.6   For a loop

Our hypnotic stunts and mind control experiments will generally be achieved by using secret gimmicks to create the illusion of real hypnosis, and applying legitimate hypnotic techniques to conceal their use. When stated like this, it may seem warped - and it is indeed (when stated like this)...

II.7   Interpretations and misremembering

Derren Brown reminded us in *Pure Effect* that the menu was not the meal, and the interpretation was not the event. What the audience ultimately remembers is even more distant that what they first understood; consequently, our set of goals should be to modify the initial perception of the effect, its interpretation, and, finally, its recollection. This last step is generally achieved by including some misremembering bits at the end of the routine (or right before the climax), when summing up the events for the spectators; the relevant tools will be described in the section pompously named "Thirteen Steps to Misremembering". These concepts are significant because reality is often created on a word-of-mouth basis, and not through factual observations.

II.8   Effects as outs

This principle is applied in most routines: the question is, how can I transform a standard trick into an out of a stronger effect? If this stronger effect fails, the out will be the standard trick we were initially considering. We will therefore offer three levels of performance:

- The stronger effect, which is presented as our primary (and only!) routine.
- The out of the stronger effect, which consists in our standard trick.
- The out of the standard trick, to be used in the worst-case scenario: this out is mainly a magicians' need (as opposed to mentalists/hypnotists' presentations) and is not always necessary, since your credibility could benefit from occasional failures.

This approach also goes against "too perfect" results: hardcore mentalists are sometimes tempted to voluntarily underplay the impact of an effect to avoid that "magic trick" feeling (for example predicting a very close result instead of a perfect match). This leeway could instead be used to risk a miracle: a success would take us to the highest level of astonishment (as per the Paul Harris scale), while failure would bring us back to the standard trick.

II.9 Risks

The universe will reward you for taking risks on its behalf.[4]

Shakti Gawain

---

[4] Please don't blame me for that one!

Risks are a direct consequence of the "Effects as outs" approach. They further confuse the spectators trying to backtrack your actions, even if you resorted to the out(s).

On the other hand, you are not jumping from a plane with a bag of overachieving silkworms: the risks are limited, since you have two parachutes, i.e. you are covered by the two outs discussed above. While you may not necessarily use them, performing with controlled risks allows for a more confident presentation, which in turn suggests that the primary effect was exactly what you intended to achieve in the first place; and this ultimately leads to a higher success rate.

II.10   Final redirections: a disturbance in the Force

At the risk of losing the few remaining readers, here is a brief abstract of a Jerome Bruner and Cecile Goodman's article called *Value and Need as Organizing Factors in Perception*:

"The problem is to understand how the process of perception is affected by other concurrent mental functions and how these functions in their turn are affected by the operation of perceptual processes. Let us, in what ensues, distinguish heuristically between two types of perceptual determinants. These we shall call autochthonous and behavioral:

- Under the former we group those properties of the nervous system, highly predictable. Given ideal "dark-room" conditions and no compelling distractions, the "average" organism responds to set physical stimuli in these relatively fixed ways.

- Under the category of behavioral determinants we group those active, adaptive functions of the organism which lead to the governance and control of all higher-level functions, including perception: the laws of learning and motivation, social needs and attitudes, and so on.

The organism exists in a world of more or less ambiguously organized sensory stimulation: what the organism sees, what is actually there perceptually represents some sort of compromise between what is presented by autochthonous processes and what is selected by behavioral ones".

Since this sort of approach focuses on the "organism" (in our case, the spectator/client) and its intrinsic characteristics, we shall add an external source of redirection (the mentalist/hypnotist) to the autochthonous and behavioral determinants. The strongest type of psychological force could be intended as a **perceptual compromise**: a synergic, interactive process combining the subliminal suggestions of the mentalist with the participant's involuntary non-verbal cues. This interaction does not lead to multiple outs, but to a feedback loop mechanism permitting flexible responses to the spectator's reactions: while occasionally widening the range of possible outcomes, it allows timely adjustments of the presentation, forcing a state of external alertness in the performer and a refreshing illusion of "free choices" in the laymen.

All this may seem a little obscure at first, but the application of these concepts in actual effects will certainly clarify the issues. With this in mind, let us begin our elitist and (hopefully) rewarding journey...

III. THE MAGICIAN'S STANDPOINT: GIMMICKS

Before starting the description of the routine, let us put aside the Words for a moment and review some of the tricky methods causing the spectators to feel heat. The following gimmicks can be used either as outs of the hypnotic suggestions (as per the "Effects as outs" approach), or as kickers to spice up a performance and reach a powerful climax.

III.1 Original version

The Hypnoheat gimmick, a.k.a. "Heat", "Hot Spot", "Hot and Cold", "Devil's Inferno", "Nate's Hot Thoughts", "Mental Heat" and "Feel the Heat", is a chemical compound based on mercury chloride ($HgCl_2$, also called mercuric chloride). It was sold in powder form, with a vial used to mix it with water, or directly in liquid form.

The original Hypnoheat trick went along these lines: before the effect, you secretly wet your fingers with the liquid substance. You then openly rubbed and folded a piece of aluminum foil (torn for example from a kitchen tin foil, or from a chewing-gum wrapper), applying the chemical in the process, and placed it on the spectator's palm. You then proceeded with some "hypnotic" suggestions to demonstrate your mental powers: the participant would feel the foil heating up and drop it after a few seconds. Alternatively, the spectator was instructed to think "Cold" and nothing happened, then think "Hot" and *voilà...* she would hastily get rid of the foil. Or, as a bar bet, a "do as I do" kind of trick where the participant was not able to keep the foil in her hand for more than a few seconds, while the magician could hold on to his piece as long as he wanted. The origins of this gimmick are unclear, but it is often credited to Jack Chanin.

For those interested in the chemical working, the heat is caused by an exothermic reaction between aluminum and mercury chloride in the presence of water. For reference purposes only, the compound was prepared with mercury, sulfuric acid, hydrochloric acid and sodium bicarbonate. Please remember that messing with dangerous chemicals is not advisable. Mercury chloride is a poison through ingestion, as well as skin contact: it is used as an antiseptic for inanimate objects and as a fungicide. In the nineteenth century, a number of milliners suffered from brain damages as a result of constantly handling mercury in their daily work – this is known as the "hatter's syndrome": *Alice in Wonderland*'s Mad Hatter character reflects this historical fact. Certainly those workers had been handling high quantities of mercury and inhaling toxic vapors every single day, for years; a drop on the thumb for a brief, occasional performance is a different matter.

Back in the '60s, the substance was quite popular among magicians: it would probably still be that way if US dealers were not afraid of possible lawsuits, such as the one that hit the magic shop "Rings and Things"... In Ulf Bolling-Borodin's

Sheherazade, Bill Palmer explains the real reason why Hypnoheat was taken off the American market: "A biker came into the Olde Magic Boutique in St. Louis (the retail outlet of Rings and Things) and wanted to know where to get the trick. He had no idea how it was done or what it involved. The people at the Olde Magic Boutique told him they did not have it. A customer told the fellow that he could purchase it from a fairly well-known magic dealer via mail order. The biker bought it and kept it in the gloves compartment of his car. While he and his 'mama' were at the beach, their child was in the car. He got thirsty, and looked for something to drink. He found the bottle containing the chemical and drank it. From what I heard, the child did not die, but he did suffer from brain damages. Even though Rings and Things had NOTHING to do with the sale of the product to the fellow in question, and even though the biker had been totally irresponsible in how he handled the chemical, the biker sued Rings and Things, and they were tied up in court for a long time over this". In addition to the mentioned book, you can find further information on the demise of Rings and Things in Jon Racherbaumer's *In a Class by Himself: the Legacy of Don Alan*.

Hypnoheat is a controversial issue: what is the real impact (on yourself and on the participant) of a tiny piece of foil briefly rubbed with the substance and held for a few seconds? Keeping in mind that cigarettes provoke cancer, is it worse than an occasional lungful of secondhand smoke? Would some professionals publicly bash the gimmick to keep it all for themselves? I solicited a number of medical doctors, and the response was quite unanimous: the real danger concerns the ingestion, and not the topical use. Besides, the performer only applies a drop of the compound on the tip of his fingers - a callous area that is limitedly absorptive. The spectator is briefly exposed to a minute amount of harmless aluminum oxide dust (on the crumpled foil). As mentioned above, mercury did cause harm in the days when it was extensively used in the make-up industry (or plainly ingested as medicine), but the performing conditions under which we operate and the small amount of substance applied do not appear critical. Without mentioning the health benefits of having your spectator actually reducing the number of cigarettes she smokes, if you use my "therapeutic" presentation.

Now, since everybody sues anybody for anything these days, the time is right for a disclaimer: I decline all responsibilities for any harm caused by any substance or method mentioned in any of my books. The effects are sold for entertainment purposes only, and shall not be made available to minors. By the way, presenting yourself as a professional hypnotist is not recommended unless you are properly trained and legally authorized: to avoid incurring in potential legal hassles, it is wise to obtain a professional certification and a performance liability insurance.

In any case, I suggest you do NOT use the original Hypnoheat because there are better options. Apart from its potential toxicity (the liquid could be absorbed into the skin), here are a few other limiting factors:

- The chemical reaction was often too fast for mentalism applications.
- It left a whitish residual dust on the foil, suggesting the use of chemicals.
- The setup was quite awkward, since you needed to carry around a bulky film canister containing cotton soaked in the liquid substance.

III.2   Updated version

Hypnoheat is now made in pellets, in a safer form. George Robinson Jr. devoted a booklet to the effect, called *Hypnoheat/Hot&Cold - The Tin Foil Trick*: it included presentational variations such as a living/dead test and a sex-appeal indicator. Ormond McGill mentioned the gimmick in the "Hypnotrix" chapter of his *New Encyclopedia of Stage Hypnosis*. More recently, Jon Tremaine performed a similar routine called "Too Hot To Handle" in his *Close-Up Mental Act* video.

The pellets are still mercury-based, and as such can be dangerous if mishandled. If you decide to use them, take a permanent marker and write "POISON" in capital letters on both sides of the Hypnoheat envelope as soon as you receive it. As with all chemicals, keep them out of the hands of children or anyone not professional enough to handle them: you may want to lock them somewhere in a high and dry place (the pellets, not the children). Always keep your fingers out of your mouth or eyes after applying the substance, and do not use the gimmick if there is a cut on the skin. Finally, make sure to thoroughly wash your hands with an appropriate detergent after the show.

Since this is a marketed item, I am not allowed to provide details about its preparation - it is much easier and safer to refer to professionals. For more information, you can contact George Robinson Jr. at Viking Magic:

P.O. Box 1778, McAllen, TX 78505 - USA
Tel: +1-956-3803929, fax: +1-956-3803930
E-mail: info@vikingmagic.com or george@vikingmagic.com
Website: http://www.vikingmagic.com

This gimmick cannot be sold in the US for the same reasons of its predecessor, but is available elsewhere. In continental Europe, you can contact my friends Flavio Desideri and Silvia Nicoletti at La Porta Magica:

Viale Etiopia 18, 00199 Roma - Italy
Tel/fax: +39-06-8601702
Email: infolpm@laportamagica.it
Website: http://www.laportamagica.it

In the UK, you can refer to Derek Lever at Taurus Magic:

32 Pilling Lane, Preesall
Poulton-Le-Fylde, Lancashire - UK FY6 0EU
Tel/fax: +44-01253-810113
E-mail: deriko@webaplomb.com
Website: http://www.taurusmagic.net

As a side note, a magic shop called "Rings and Things 2" recently opened in Virginia - but I suspect they don't carry the gimmick.

An envelope containing about 10 pellets costs around 20 US$, and each pellet will last for approximately 20 performances. For the mathematically oriented, this corresponds to a cost of about 10 cents per performance. The pellets are extremely thin and less than one inch long[5] (Figure 1, shown close to a cigarette) - you certainly realize that bulkiness will never be a problem.

---

[5] 1 inch=2.54 cm.

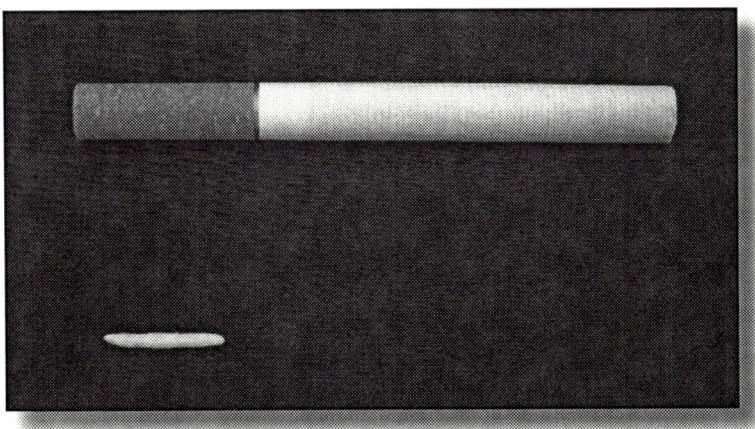

Figure 1

Let us briefly describe the proper way of using this gimmick. Tear a small piece of tin foil (less than one-half square inch). Moisten your left first finger and thumb, for example by casually bringing them to your mouth (I usually pretend to stroke my beard). This is done BEFORE applying the chemical, of course: NEVER put your fingers in your (or somebody else's) mouth or other orifices AFTER having touched a chemical substance. This wetting action is not always necessary, but in cold and dry surroundings a slight humidity facilitates the reaction. Rub your left thumb and index finger on the pellet in order to obtain a slight (and virtually invisible) whitish residual dust. The substance will stay on your fingers for about 10 minutes or more, if you make sure not to touch anything before the show (this is why right-handed performers should apply the chemical on their left fingers): this allows you to set up well in advance. Gently rub the foil for about 5 seconds, fold it in half, rub it again for another 5 seconds, fold it in half one more time and quickly roll it into a tiny ball. For our purposes, the foil should be as small as possible: you will need to complete the fold using your nails, and have your second finger assist in the process. Finally, place the foil in your right hand (Figure 2); after about 10 or 20 seconds, you will feel a sudden burning sensation forcing you to drop it.

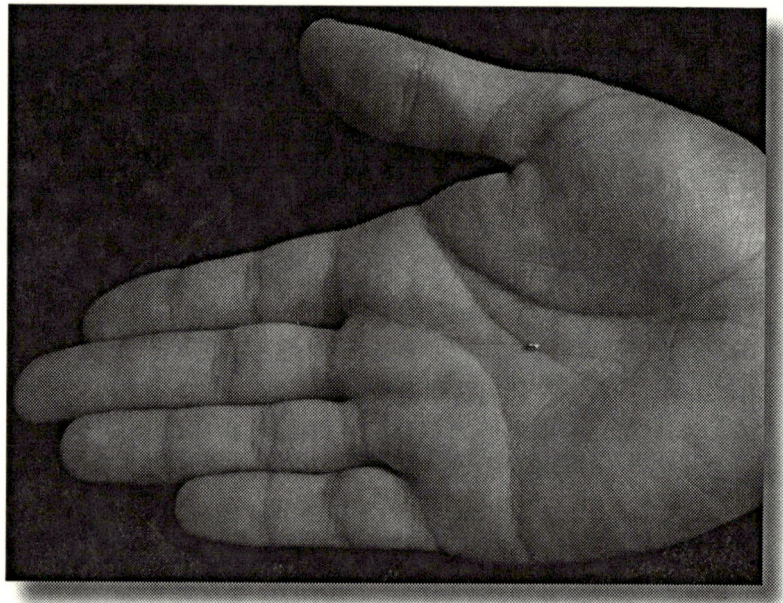

Figure 2

If you are using a piece of foil taken from a cigarettes pack or a chewing-gum wrapper, please note that this foil is actually made of one layer of aluminum and one of regular paper (Figure 3, showing the two separated layers): during the rubbing/folding process, make sure to keep the shiny aluminum side on the outside, since it is the one reacting with the chemical. Do not apply too much pressure, or you might separate the two layers, hampering the reaction.

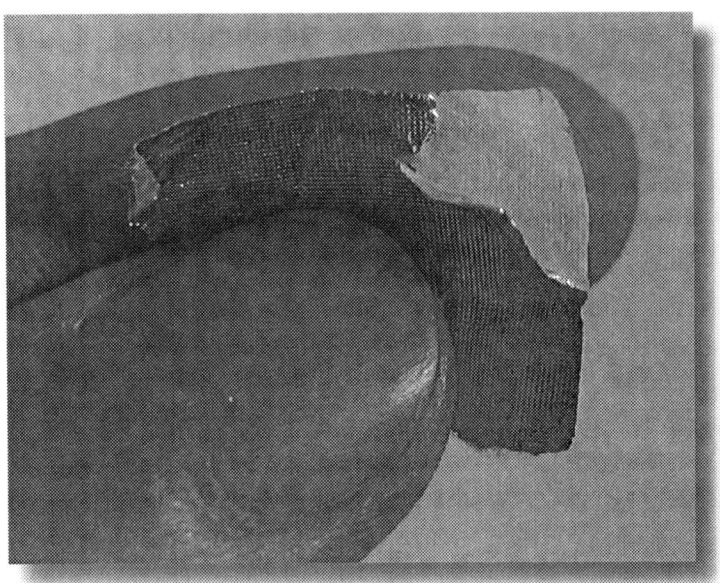

Figure 3

A great quality of this gimmick, apart from being hardly seen nowadays, is its "hot and cold" feature: the spectator won't feel anything for a while, until a sharp and unexpected burst of heat will force her to drop the foil. If she picks it up afterward, its temperature will have returned to normal, leaving her clueless.

Still, the Hypnoheat basic presentation creates a challenge, a "magician versus audience" situation where macho spectators often deny the actual heating sensation in their hand, to prove their mental strength against the performer's hypnotic suggestions. The effect in my views had to become an out: I first started to hide the rubbing movement, continued by minimizing the perception of the folding process, and ended up in eliminating (or at least concealing) the foil itself.

III.3 Progressive variation

This version, sometimes called "Hypnoheat 2", is 100% politically correct. Even Max Maven was unable to track down its originator, due to the peculiar nature of the gimmick: it's the wrapper of a cinnamon-flavored gum called "Wrigley's Big Red".

Tear a piece of the wrapper, moisten the internal coating (not the shiny aluminum side) and place it on your palm, wet side down: the cinnamon coating will react with water and become warm after about one minute. In theory, the sweat on the palm may be enough to start the chemical reaction, without secretly humidifying the paper – this means that the trick could be performed without you ever touching the wrapper: you would simply give out your own pack of non cinnamon-flavored gums, having previously substituted the wrappers with the ones coming from a Big Red pack, and ask the participant to take a gum, tear a piece of its wrapper and hold on to it. However, I had limited success with this strategy since its reliability mainly depends on the right weather conditions (high temperature and/or high humidity). Alternatively, you could borrow a gum, tear a piece of its wrapper and switch it with your moistened cinnamon foil: you can do it easily because most chewing-gum wrappers look exactly the same.

The advantages of this variation, compared to the Hypnoheat method, are that you do not need any chemical gimmick, and no folding is required. On the other hand, the timing is different: the paper will get warm after about one minute and gradually become hotter – as opposed to the sharp, unexpected burst of heat obtained with the Hypnoheat gimmick. You will also need a larger piece of foil, which is harder to conceal.

Concerning its availability, the "Big Red" is widely distributed in the US but not marketed in Europe (who knows what a kid did over there...). You can contact the manufacturer and ask if there is a reseller in your country: the Wrigley company was founded more than a century ago and has quite an international presence. The Headquarters' address is:

Wrigley Jr. Company
P.O. Box 3900, Peoria, IL 61614 - USA
Website: http://www.wrigley.com
(it includes the contact details for a number of countries).

You can also find the gums in online candy shops such as:

Cybercandy
11 Shelton Street, Covent Garden,
London - UK WC2H 9JN
Tel: +44-0207-2405505, fax: +44-0208-8018815
Email: enquiries@cybercandy.co.uk
Website: http://www.cybercandy.co.uk

A 5-piece pack sells for about 1.5 US$[6]; once again, the mathematically oriented will joyfully realize that, considering 3 performances per gum, the trick has the same shivering cost of 10 cents per performance.

This method does not work with all brands of cinnamon-flavored gums. Before wasting your time in experimenting new solutions, pun intended, I have also tested diluted cinnamon

---

[6] If you resist the temptation to purchase something else to justify the shipment costs... Check out your local store before ordering anything!

powders and essential oils (used in baking) – they had no satisfactory effects. On the other hand, cinnamon-based products come in different concentrations: perhaps you will be luckier and obtain better results... I did not experiment further, because those oils and powders leave a noticeable smell that would certainly cause suspicion.

Concerning the safety of this gimmick, there does not seem to be any real threat: gums are meant to be eaten, after all – even if they may not be the healthiest food on the market. In rare allergic cases, the hot wrapper could leave a small temporary mark on tender skins, especially if you use water to wipe the affected area after the trick (...)

III.4 Marketed props

Several marketed props adopt a methodology similar to the original Hypnoheat concept, but the chemicals are safely sealed on the inside. For example, the "Hot Ball" is a metallic sphere about the size of a tennis ball that gets hot in the spectator's hands. Those items are also made in the form of teakettles (an homage to the Great Magic Kettle acts of the nineteenth century), coffee boilers etc.

Of course, you end up with a magic prop.

IV. PERFORMANCE

This is the effect I have been successfully using for years as an opener: it elicits strong reactions in the audience due to the importance of the topic and the positive suggestions embroidered into the script. Since it is presented as a hypnotic experiment, the complete routine is a bit long (almost 10 minutes), but it can be shortened to about half its length; we will first describe the full original act with all its outs (including the conclusive chemical kicker), before analyzing a few variations around the same theme.

IV.1 Optional preparation

The best gimmick for the out/kicker of this effect is the Hypnoheat pellet. Before proceeding any further, you may want to reread the disclaimer in the copyright page: "Anyone using any of the techniques, gimmicks and ideas described in this book does so entirely at his or her own risk: the author and the publisher cannot be held liable for any loss, damage or injury whether caused or suffered by such use or otherwise". This is valid especially (but not only) for readers living in the US, where one TV ad out of three offers lawyer services.

If you plan to use it, the preparation is as follows: before leaving your home, empty your left trousers pocket and clean it thoroughly. I generally use rolled tape to remove any residual dirt. Now slip one pellet inside, along with a tiny piece of kitchen tin foil. If you don't have a pocket separator, it is wise to stick a bit of double-sided tape inside the pocket, and press one end of the foil against it: this will allow you to locate it without fumbling later on, and will prevent a premature contact with the chemical. The tape is stuck close to the top-right corner of the pocket (Figure 4, showing the left pocket inside out): to find out its exact position, simply put your left hand in your pocket and pinch the fabric with your thumb and first finger, where it is more comfortable for you.

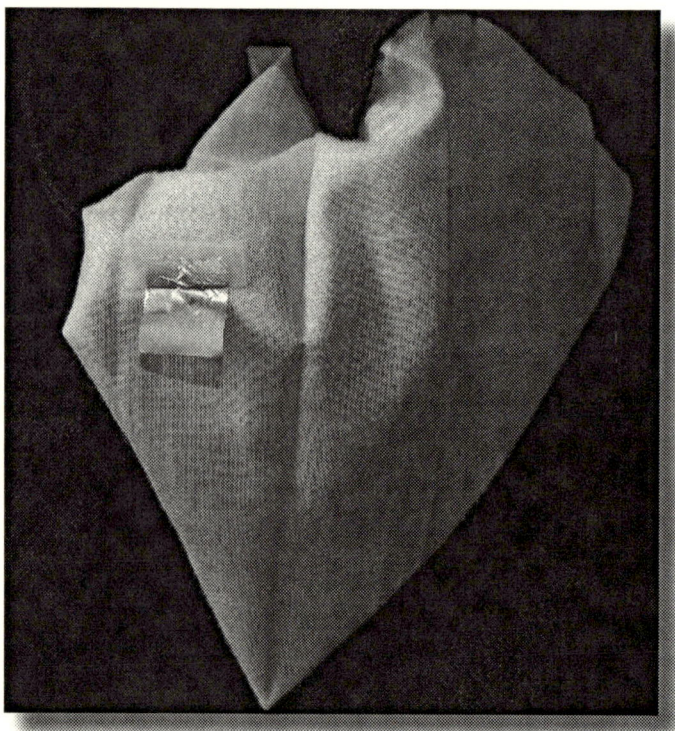

Figure 4

There is no need to apply the chemical in advance since you will be able to set up under fire, so to speak, and only when necessary.

IV.2   Opening: a gift in season

Henning Nelms in *Magic and Showmanship* stressed the importance of performing the right effect at the right moment: magically producing a ham sandwich when a hungry spectator is requesting for it makes the trick one thousand times more powerful. In informal situations, the routine should be performed when the conversation naturally drifts toward smoking (of course nobody prevents you from subtly forcing the issue...). On the other hand, when specifically requested to "show them something", my opening line is the following:

> *How many smokers do we have here... Could smokers please raise their hand* [raise hand] *for a second? Any of you kids?*

This line involves both smokers and non-smokers: the latter usually participate too, asking their smoking friends to raise their hands. This request makes your act lively right from the start (compare it with a dull "I want a smoker"), and lets you spot the most responsive spectators - the ones you may be involving in forthcoming effects.

Concerning the movements, the miming action accompanying the request allows you to casually show your hands empty, without of course mentioning it. It is wise to start a show with empty hands: this is when the spectators are the most suspicious, and scrutinize every single move of the performer.

Now, I would hate to start bothering you with outs even before the beginning of the effect, but what if we can't find a smoker in the audience? We would continue the routine by reframing the presentation around a *prevention* experiment for non-smokers, instead of a *cure* for smokers. This variation is described in the relevant chapter: there will only be a few words to change. Alternatively, feel free to perform any other killer impromptu effect targeting non-smokers! For now, let us proceed with the planned presentation, assuming that there is at least one smoker in the audience.

IV.3   Selection of spectator

When possible, select among the responsive smokers a nicely-dressed young woman. This choice is not only a personal preference, but is justified by the following reasons:

- There is a common tendency to listen to what a young and pretty woman says, and disregard what the wise old man is suggesting.
- Women are usually dressed better than men, and are nicer for the audience to look at - offering greater misdirection opportunities.
- They are more likely to play along than their strong and silent counterpart.
- Burns and scars are generally more frightening for women; most of them are not ashamed of showing fear, and sometimes overreact.

In any case, when selecting a participant, keep in mind Benjamin Franklin's sound advice ("Don't trust the young doctor and the old barber") and my two cents: beware of smiling faces.

IV.4 How names hold the key

Let us call our spectator Sue, for example.

Sue, how did you start smoking?

This open-ended question builds rapport without being corny. Asking "Do you remember your first cigarette?" could be perceived as trying to pull her heartstrings, similarly to some first kiss / dead uncle lines. "How did you start smoking" sounds more generic, yet elicits the same reactions and triggers the same memories: by recalling how she started smoking, she will in fact remember her first cigarette, the sensations she felt, the friends she was meeting in that period etc. Besides, asking "*How* did you start smoking" instead of "*When* did you start smoking" leaves room for a number of interesting answers.

An important linguistic tool: never pronounce the spectator's name in vain... It is such a powerful keyword, that you want to keep it for those special moments: I use it for marking purposes and for establishing eye contact. For example, imagine walking in a crowded market, among hundreds of noisy vendors trying to pitch you... you will irremediably turn toward the one who calls your name. Mentioning her name at the beginning of the patter was not strictly necessary, but it builds rapport and helps you remembering it.

IV.5 Rapport

My father used to smoke too, but I slowly made him STOP [downward gesture]

I am contributing a personal experience as well, which will help gaining rapport and have her reciprocally share her inner feelings. This anecdote is (mostly) true for me; besides, it easily links to our previous question, implying HER relationship with her father, her parents' reaction when they learnt she had started smoking, the difficulty in breaking the smoking habit, the need of being supported by close relatives, and much more... Sharing a personal experience is often an effective way of having the participant reciprocally open up. Of course substitute the word "father" with auntie, son etc. at your convenience - possibly something as close as possible to the so-called truth.

On the word "stop", gesture downward with your hands, palms down and fingers apart, in a typical relaxation request. Never miss a chance of casually showing your hands empty! We shall refer to this movement throughout the routine as the "downward gesture".

IV.6   Yes Set

In informal situations, somebody will usually enquire about the method you used. Continue in any case with:

I can show you how I did it, if you think you may try to STOP too one day, yes?

We need a positive answer here, to begin our experiment. Ideally, all we should hear from now on is "Yes". Maybe the spectator doesn't want to stop smoking right now, but she certainly *thinks* she *may try* to stop, *one day:* we have strongly qualified our question, because a straightforward "Do you want to stop smoking?" may have brought us a dry "No" (or

"Not now") in return. The qualifiers "slowly" ("I *slowly* made him stop") and "one day" ("you may try to stop too *one day*") help indicating that this is not a procedure with immediate results, but a slow process that will realistically bear positive effects in the long term.

An affirmative reply also means that she recognizes her addiction (and acknowledging the existence of a problem is always the first step toward solving it), and that she considers the possibility of quitting smoking.

IV.7   All in her hands

So please bring out your cigarettes [grin]

This brutal, unjustified and unembellished request will slightly worry the spectator[7]: "This weird person wants to make me stop smoking, and asks for my precious cigarettes with a devilish grin? What for? Will I get them back?"

Her natural hesitation provides an excellent excuse to continue with:

Oh, I'm not going to touch anything [downward gesture], *in fact I'M NOT GOING TO DO ANYTHING AT ALL* [hands in pockets], *EVERYTHING WILL BE IN YOUR HANDS!*

The statement of intent couldn't be clearer: "Everything will be in your hands" not only reinforces the notion that you will never

---

[7] Especially if you vanished her lighted cigarette beforehand... The last thing she wants is to give you the whole pack!

touch anything, but also begins to suggest that the success of the experiment only depends on her. The sentence "I'm not going to do anything at all" is further highlighted by the action of innocently placing your hands in your trousers pockets. You obviously can't do anything in this position, can't you? This allows you to rub your left fingers on the pellet, if you plan to perform the chemical kicker and did not have the chance to set up in advance. The routine will offer another opportunity later on, letting you decide on whether to use the gimmick or not (depending on the outcomes of the suggestions). The minimal movements of your left fingers inside your pocket will pass completely unnoticed. If you wear a jacket, you could benefit from an additional subtlety described by Mark Strivings in *Miracles from the Hip*: put your hands in your trousers pockets so that the bottom sides of your jacket will cover them (Figure 5), further concealing the movements of your fingers inside your pockets.

You may have noticed that we are not borrowing or taking anything: she is only asked to *bring out* her cigarettes, not to give them to you. Please compare this line with a plain "Could I borrow your pack of cigarettes?". Once again, every word and every gesture throughout the routine must suggest that everything will be in her hands, and that you will never touch anything at all. We don't even mention the "pack" of cigarettes: since it has a covert role in this version of the effect, we only refer to the "cigarettes". The presence of the pack is irrelevant: it is borrowed (hence ungimmicked), and will be left in plain view, untouched.

Figure 5

IV.8 Framing

So, let me show you this brief hypnotic [pause] *EXPERIMENT used to cure smoke addiction.*

Instead of "hypnotic experiment", you could of course talk about psychological cure, magic trick, esoteric ritual etc. depending on your presentational preferences. However, do not talk about a *test*, as it would make our participant nervous and not properly responsive: an experiment can become a test only after its successful completion. We labeled the experiment as "brief", because she may confuse it with some kind of hypnotherapy session, and may not be willing to grant us so much time.

If you are a smoker, somebody may ask why this cure was not used on you; the excuse could be that the method is not applicable to a subject who already knows its modus operandi.

IV.9   On her side

THIS EXPERIMENT WORKS, but only under the right conditions: everything depends on your intention to cooperate, and I really want you to be successful...

"This experiment works" is marked by direct eye contact with the participant and a stronger (or softer) tonality: IT WORKS, you pulled it off in the past with your father, auntie, son etc. Still, the word "experiment" presupposes a degree of uncertainty and a possibility of failure; in fact, it only works "under the right conditions". This brings the audience attention on whether the participant will succeed, not on *how* the experiment will be achieved. We also remain unspecific about those "right conditions" - this leaves us with ample relabeling possibilities if things go wrong.

"Everything depends on your intention to cooperate", similarly to "Everything will be in your hands", helps implanting the idea that the experiment is under her control (and, consequently, that you won't do anything tricky). It is also a nice way to share responsibilities, suggesting that she may be partially blamed in case of failure. The line "I really want you to be successful" comes deep from the heart, since a lack of collaboration from her side would certainly spoil the effect... Uri Geller and many others after him often used this approach: gently taking the participant on your side, building rapport and giving her all the credits in case of success. This is a complete departure from the original Hypnoheat presentation, where the performer tried to influence the spectator against her will, in a challenging manner.

IV.10   Body positioning

... [to the audience] *so please let her relax* - [to the spectator] *would you mind moving over there* - *so we can clearly see what's going to happen.*

The patter concisely provides two valid reasons for the repositioning of the spectator:

- The audience will have a better view of the event ("so we can clearly see what's going to happen"[8]). Mentalism and stage hypnosis aren't visual arts, and the spectators' reactions are sometimes all we have; Marc Salem, of *Mindgames* fame, once referred to them as "living props". Make sure that the

---

[8] Presupposing *en passant* that something is actually going to happen...

participant can be clearly seen by everyone at all times, by indicating where you want her to stand in a discrete and functional way.

- Stepping a little bit away from the rest of the audience will supposedly help her to relax and focus on the experiment ("please let her relax").

The spectator is simply positioned in front of the audience, at a distance of about 10 feet (3 meters), depending on the performance settings and the lighting conditions. This distance is required only if you plan to use the chemical out/kicker. You should stand halfway between the audience and the participant: if there isn't enough space in the room, it is best to stay closer to the audience, non-verbally suggesting, once again, that you will never do anything at all. You should be facing the participant and stand slightly on her right. This positioning further covers your left side from the audience view, concealing the activity of your fingers in your trousers pocket, and will minimize the perception of your forthcoming motion: during the effect, you will in fact get closer to the spectator in a left to right, clockwise direction (from the audience point of view), execute a covert move (if necessary) and naturally step away in a continuous movement (Figure 6). Gary Kurtz described this odd notion in *Leading with your Head*: a left to right (and, may I add, clockwise) motion is generally perceived as more natural and less eye-catching than a movement in the opposite direction. It probably has to do with the way we read[9].

---

[9] An overkill note: if performing for Arabic audiences, who read from right to left, you would need to invert those positions.

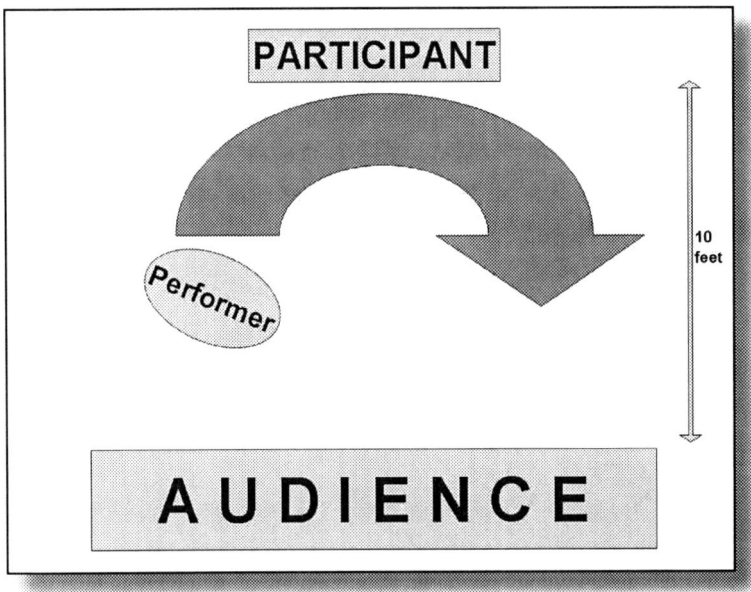

Figure 6

In addition to the physical positioning, we verbally distance ourselves from the spectator by asking her to move *over there*, not *here*: this suggests a greater distance. Besides, we shift the reference points by placing ourselves with the audience ("so *we* can clearly see what's going to happen"): "We" (the performer and the rest of the audience), as opposed to "I" (the performer) or "They" (the rest of the audience).

IV.11 Arm/hand positioning

You need to hold your right hand [arm positioning] ... *okay Sue, and RELAXed* [hand positioning]

Yes, it doesn't read too well but this part is based on visual cues. Your left hand remains in your pocket, unable to do anything tricky, while the right is brought out to illustrate the arm and hand positions: since I am right-handed, I generally use my left hand for the covert moves (at least the ones not too technically demanding), because it might be less suspected than the right. We ask the participant to use her right hand as well for consistency with our miming action and for misdirection purposes, to distract the audience from the *position* of the hand, which is the real deal. Let us briefly describe the exact positioning:

The spectator's right arm should be almost outstretched (Figure 7). You provide indications on how she should hold it by extending your arm, starting from a semi-curved position and fully stretching it. This position is justified by the following reasons:

- It maximizes the visibility of her hand, independently from the performance angles, and suggests openness and fairness. The most experienced readers would know that this rule is usually applied to performers, when holding their own props...
- Her arm will naturally get tired when kept out stretched, making her feel some of the sensations described later. She won't even be able to hold it with her other hand, because it will soon be busy with an imaginary cigarette.

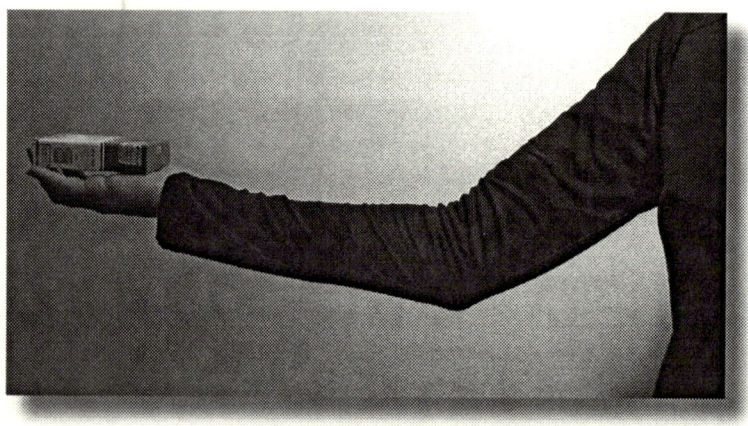

Figure 7

Her right hand should be palm up and relaxed, in a receptive, open position. This ashtray-like pose is necessary for the chemical out/kicker: if you used the gimmick for any length of time, you certainly found out at your expense that it does NOT work with the same intensity in a closed fist, or when the foil is completely covered. Therefore, if the spectator was holding her hand wide open, i.e. with straight and tense fingers, the pack of cigarettes would lay flat on her palm, covering the foil and hampering the heating process. A relaxed hand has instead slightly curled fingers, which create a gap between the palm and the pack (Figure 8) and ensure that the chemical will react properly.

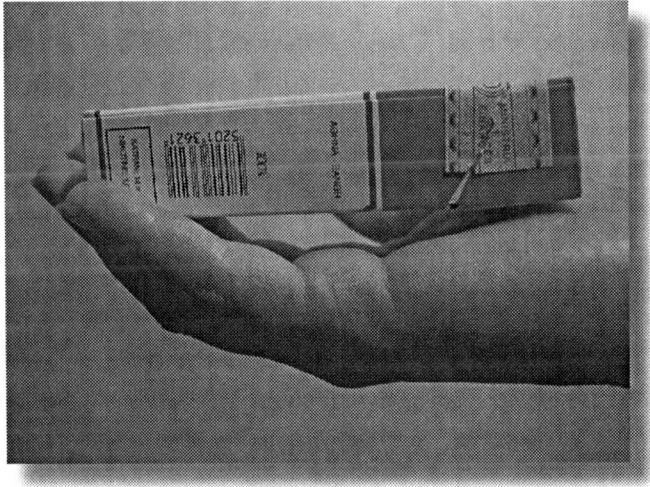

Figure 8

Since this position is such a critical issue, we verbally minimize its importance by barely mentioning it, and by providing a visual cue instead: just as before, casually show the spectator how to hold her hand by curling your fingers (exaggerating the movement). To make sure that she picks up the cue, you should first establish eye contact when calling her name

("...okay, *Sue*, and..."), then drag her look toward your right hand by shifting your gaze from her eyes to your fingers, and finish the sentence ("...RELAXed") while curling your fingers.

There is a harmless dual-reality situation here. The curling gesture is made with your hand at stomach level and close to your body, in order to hide it as much as possible from the rest of the audience: most spectators will only understand that she has to relax (consistently with your previous requests), while the covert gesture you make indicates to the participant that *her fingers* need to be relaxed (that is, naturally curled) as well.

We are loyal to the John Carroll and Michael Tanenhaus' "Minimax" principle: pure economy of language and motion, minimizing the patter and the gestures (hence, keeping the audience attention) while maximizing the communicative results. We also reduced the number of clues given to the audience, by concealing some of the movements from their view.

One last note about the linguistics: once again, we are not mentioning the pack of cigarettes. We didn't say "You need to hold *the pack* in your right hand", but simply "You need to hold your right hand", without any further reference. The way she positions her hand will obviously determine how the pack will be held.

IV.12 Spectator conditioning

It may hurt a little bit but won't leave any physical evidence in the long term. Even though, some people felt REAL PAIN [burning gesture]

Painful skin burns and ugly scars are generally quite scary (do not push it too far, or the participant may refuse to continue the experiment at this early stage!). We said that it would not leave any physical evidence "in the long term", but what about scars in the short or medium term? And how long is this long term anyway? Besides, some unspecified people felt REAL PAIN... pretty frightening thoughts. We obviously qualified the evidence as "physical": the experiment won't leave any *physical* evidence, but will hopefully bear psychological consequences – after all, this is the main purpose of hypnosis.

The pain is visually indicated by a miming action we will be calling the "burning gesture": on "real pain" ("even though, some people felt *real pain*"), hastily reverse your right hand as if you had felt a sting on your palm. This begins to condition the participant to what is going to happen and how she will need to react. It is best to hide this movement from the rest of the audience by covering it once again with your body, to avoid an anticipated revelation of the climax: the burning sensation and the possible fall of the pack should remain a secret for the surprising finale.

IV.13   Seeding in her shoes

> *It seems you're getting a bit* [smile]... *but don't worry, when you feel any pain just STOP* [burning gesture] *AND YOU'LL BE FINE, okay?*

The line "It seems you're getting a bit..." is used to build rapport: you put yourself in her shoes, and claim to understand that she may now feel a bit nervous due to your previous warnings. However, you are not specifying the exact feeling, but leaving it subtexted, with a smile: the spectator and the rest of the audience will get ahead of you, filling the blank and identifying with whatever sensations they think THEY would feel in that situation. This subtlety has three main objectives:

- It increases the audience participation by allowing them to complete the sentence with the sensations they fancy the most.
- It shortens the patter, since we don't have to provide further details.
- It may suggest that we know exactly what they feel.

The burning gesture is repeated on "stop" ("when you feel any pain just *stop*"), and its meaning is very clear to the participant: when she feels any pain (leaving it very unspecific about the type of pain, but presupposing that she will indeed feel it), she has to stop the experiment and reverse her hand. Once again, we are conditioning her to drop the pack when she decides to stop. Another seed planted for future harvests... We warned her, everything is in her hands: the choice between being left with painful physical evidences or a velvet palm is only hers. This will help amplifying her forthcoming sensations and make her drop the pack at the first hints of pain. The burning gesture is once again covered by your body, and unseen by the rest of the audience. Most of the spectators only understand the words you say, that is: when she feels any pain, she can decide to stop the experiment - with no visible or verbal reference to the pack.

You may have noticed the subtext and the relevant positive suggestions: dropping the pack could be seen as a metaphor for getting rid of the cigarettes, i.e. for stopping smoking (not just stopping the experiment). And stopping will make her feel better ("stop and you'll be fine").

We end this brief seeding patter with a rhetorical "okay?" bound to obtain another affirmative answer – it gives the illusion of a question, hence of a choice. This is the second "Yes" in a row (or third, if she replied affirmatively to our request of bringing out her cigarettes): we are also conditioning her to answer positively, because harder questions are about to come.

IV. 14   Pre-trance fact

So, you're standing over there, still nice and relaxed, right? [Smile]

It is advisable to start the induction with something easily observable, before moving into the questionable "trance" phase. She is indeed standing over there[10]: this fact cannot be refuted, and allows us to obtain another easy "Yes".

You certainly noticed the embedded suggestion about her being nice and relaxed (and "right"). However, she is *still* nice and relaxed, presupposing that she might not feel that way later: this could be due to that painful sensation expected during the experiment, or, on a deeper level, to the fact that smoking will take its toll soon or late...

IV. 15   Induction

Good, now could you please take out an INVISIBLE cigarette [pinch grip, mime], and as you stand there smiling, you can begin to relax deeper and try to visualize it in your hand... Sue, when people CONCENTRATE [downward gesture, close eyes] *or*

---

[10] Please remember that she is standing "over there", i.e. far away from us, not "here".

focus they tend to visualize much better... and when you visualize that white, long cigarette softly resting between your fingers [pinch grip] *you can notice its light weight and s l o w l y take it to your mouth now* [mime], *feeling the harder texture of the brown filter against your lips... now let's light it* [mime]... *and when smelling that gray smoke, the most sensitive persons almost feel* [inhale noisily] *the full taste* [exhale noisily] *of tobacco...*

This patter involves the participant's five senses: she visualizes the white, long cigarette, feels its weight and the texture of its filter, hears you inhaling and exhaling noisily, smells the smoke and tastes the tobacco... We will ask for confirmation later. Create vivid mental images and leave her enough time to actually experience those sensations. You will find out that the rest of the audience (even the non-smokers) will also get involved.

On "Now let's light it", step forward and act as if lighting her cigarette with an imaginary lighter. This humorous move is a great excuse to get a little closer: since her hands are busy (with the pack and the invisible cigarette), you gallantly pretend to light it for her. Do it at the fingertips of the right hand, arm extended and without touching anything. Immediately step back afterward, but without returning to your initial position: you are now closer to the spectator, still on her right side but almost at arm's reach. This gradual approach is preferable because standing still throughout the entire routine and suddenly moving toward the participant during the tricky moment (at the end of the effect) would draw much more attention. This gag would be out of place in a serious hypnotism session, but I prefer to play it lighter because humor is disarming.

Some additional notes about the linguistics used in this paragraph:

- Make sure to underline that she is supposed to take out an INVISIBLE cigarette, because inattentive participants sometimes reach for a real one.
- Stack your patter in order to avoid any interruption, and gradually slow down the pace, using a softer and deeper tone and a longer breath patter (indirectly suggesting a relaxation process).
- A suggestion is usually more powerful when linked to something unquestionably true and immediately observable (even without a logical connection). For example, we said: "As you stand there smiling (fact), you can begin to relax deeper (suggestion)". Or, you can read this sentence and notice how clearer it is becoming for you now. For the enthusiasts, those are called "adjunctive suggestions".
- We carefully avoid direct orders: people do not fancy them and prefer subtle, motivated suggestions. Most of the commands are embedded, for example "When people concentrate or focus, they tend to visualize much better" instead of "Concentrate and visualize", or "The most sensitive persons almost feel the full taste of tobacco" instead of "Feel the full taste of tobacco". At the same time, legitimate reasons are provided for each command: for example, by concentrating or focusing she will "visualize much better" and consequently increase her odds of succeeding in the experiment. The interrogative inflection given to some of our commands ("Could you please take out an invisible cigarette") also suggests the possibility of choices: they look more like questions than orders.

If you really have to ask something directly, at least don't forget the magic word "Please"... No artist ever got booed by overestimating the sensibility of the audience.

- When pretending to light her cigarette, we said "Let's light it", not "Let *me* light it", unilaterally sharing the responsibilities.
- She has to visualize *that* cigarette in her hand (and smell *that* gray smoke): you are not referring to *this* cigarette right here, but *that* cigarette over there - further distancing yourself from the participant.

Finally, a few comments about the physical subtleties:

- Move slowly (consistently with your tonality and breath patter), indirectly suggesting to the spectator to do likewise.
- The "pinch grip", well known to magicians, refers to the position of your fingers when pretending to hold the imaginary object: the invisible lighter (or cigarette) is seemingly held between your right thumb and index finger (Figure 9). After the miming actions, naturally separate your fingers, indirectly showing your hand empty. This gesture will be employed in a tricky way at the end of the effect, when your left thumb and index finger will secretly hold the tiny piece of foil: it is consequently introduced well ahead of time and used without dupery throughout the routine, so that the audience becomes accustomed to it. This is a useful subtlety in real close-up settings, when spectators are burning your hands (sorry I couldn't resist anymore).

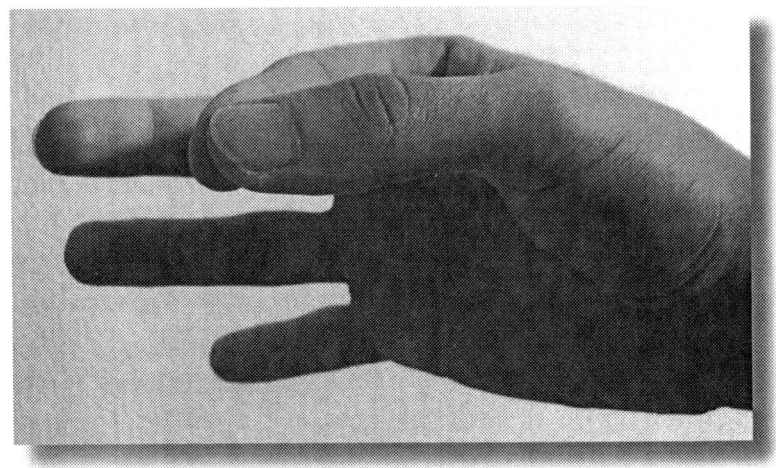

Figure 9

IV.16 Acting

Make sure that the participant acts congruently as well by opening the pack before removing her invisible cigarette. Depending on the type of packaging, she might leave it slightly opened: this would allow some of the cigarettes to come out during the fall of the pack, for an even more visual climax.

Some additional humorous possibilities:

- On "Take it to your mouth now", you could pretend to bring your own invisible cigarette to your lips, and pronounce the next sentence ("Feeling the harder texture of the brown filter") in a distorted way, just like if you had a real cigarette in your mouth. You would then apparently remove it and speak normally again.

- Instead of a lighter, you could use an invisible box of matches. The miming action, done with both hands, is more visual and provides another gag opportunity: after supposedly lighting her cigarette, you could pretend to blow off the imaginary match, have a look around as if asking yourself "how am I going to get rid of this one" and throw it discreetly in a funny place around you (a vase, a drinking glass etc. - anything as long as you don't move around). You could then replace the invisible box in your left trousers pocket, and take advantage of this new opportunity to set up ("Never waste a trip to your pocket", to quote Al Goshman).
- An alternative excuse to move toward the spectator could be to bring her an invisible ashtray, later in the show.

These gags obviously depend on your performing style and on the presence of children in the audience.

IV.17   Police and thieves

Why wouldn't you let the participant take out a real cigarette? We use an invisible one for the following reasons:

- A real lighted cigarette may truly burn her, due to falling ashes or to the proximity of the lighted end to her skin. There would be nothing magical or hypnotic about that!
- We obviously don't want to worsen her health conditions by asking her to smoke a real cigarette.
- Nowadays, reaching for a cigarette is perceived as an obscenity in an increasing number of places.

- It is much faster to deal with imaginary items than to actually remove a real cigarette, look for a lighter etc. – we want to maintain a quick pace and keep the audience attention.
- Invisibility offers greater entertainment possibilities and catches the spectators' imagination.
- This approach eliminates one unnecessary "prop" from the audience visual (remember that we want to introduce as few elements as possible); we used an invisible lighter for consistency.

So, a heckler may wonder now, why did she bring out a real pack if she doesn't take a real cigarette out of it? In this version of the effect, the pack will be used to cover the foil you will surreptitiously drop in her hand: this justification however would not be well accepted by the audience. Keeping in mind that we never asked explicitly for the "pack", a rational reason is provided for its presence: she needs the pack to remove a cigarette, that's what a pack of cigarettes is for after all. True, it's just an *invisible* cigarette, but we don't want her to use a real one for the reasons explained above. There is a weird internal logic that has never been questioned. Please note that we refer to the cigarette as "invisible", not "imaginary": it is actually there in the pack (and she should take it out), the only difference with a real one is that it cannot be seen...

The presence of the pack will be further justified in the "Old Cowboy's last trick" paragraph. In any case, it is not necessary to explain why she needs to use an invisible cigarette instead of a real one, or why she should extract it from a real pack: it would only slow down the pace without bringing any additional benefit. Provide excuses upon request only – there is no need to run if you are not being chased.

IV.18   Trust and responsiveness

Sometimes the spectator will voluntarily close her eyes at the beginning of the induction, to better visualize the scene. This indicates a particularly receptive subject. You may want to force the issue and throw in an extra deception, evaluating at the same time the responsiveness of the participant: establish eye contact when calling her name, and close your eyes for a few seconds on "concentrate" ("Sue, when people *concentrate* or focus...") as to better feel the suggested sensations. Chances are high that she will close her eyes as well. The reason provided is that concentration (i.e. closed eyes) helps in the visualization process ("... they tend to visualize much better"). Of course we never ask her directly to close her eyes, and the rest of the audience does not see our visual cue. Incidentally, "When people concentrate or focus" is a mental magician's choice: we graciously offer her two identical options. Giving the illusion of free choices is always refreshing, and makes the participant focus on *how* she is going to comply, not *if* she will do it.

This subtlety is useful to evaluate the responsiveness of the spectator, in case you were planning to involve her in riskier effects later in the show. Besides, it makes room for an extra deception. After she closed her eyes, continue for a moment with the patter and add the following sentence before pretending to light her cigarette:

> *Of course people usually begin to feel a bit sleepy or heavy-headed but please try to stay awake and relaxed, okay?*

This suggests to the other spectators (who are not aware of our visual cue) that she might already be falling into some kind of trance. The additional "Yes" secured in this occasion may in fact confirm this impression! We said "Of course people usually begin to feel a bit sleepy or heavy-headed", because everybody knows that during hypnotic experiments, people may fall into a state of trance and feel this type of sensation...

The equivalence concentrate/close eyes will be necessary in the "Final out" phase, when the participant will really need to close her eyes for a moment. Since you might resort to this out at the end of the effect, it is wise to introduce the association well in advance. For now, we subtly invite her to reopen her eyes ("Please try to stay awake") but, once again, without asking it directly. This effect can work with the spectator staring at you at close distance, so why miss the chance to appear fair and open? If she keeps her eyes closed, leave this wonderful, trustful creature in her apparent trance state: she will naturally "wake up" later.

IV.19   The almost sixth sense

... and you can almost sense its smell, can't you?
[Silence]

This question is a continuation of the sentence "The most sensitive persons almost feel the full taste of tobacco..." described in the induction paragraph (I thought I would remind you of it, since many pages of explanations separate us from that line); however, it deserves its own paragraph.

In smoking environments, where you can naturally smell tobacco in the air when inhaling deeply, this sensation will be easily acknowledged by the spectator. In non-smoking settings however, you could use this additional convincer: before leaving your home, place some tobacco in your right trousers pocket (simply remove it from the tip of a cigarette). Rub your right fingers on it after placing your hands in your pockets, on "I'm not going to do anything at all" (in the "All in her hands" paragraph). When pretending to light her invisible cigarette, your right fingers naturally move very close to her nose: immediately continue with the induction patter ("The most sensitive persons almost feel the full taste of tobacco") and step back during the noisy inhalation cue: this is when she will inhale deeper as well and smell the tobacco. You are away when it happens, and due to the barely detectable odor and the distance from the rest of the audience, she will be the only one smelling it. Try this convincer, it works wonderfully!

In addition to this tricky bonus, we use the following supporting tools to make sure that the participant agrees:

- The qualifiers: perhaps she cannot strictly smell and taste the tobacco, but she can *almost sense* it. "Sense" is rather unspecific, and has this vaguely psychic flavor.
- The way the question is phrased: we use an affirmation coupled with a negative question. We are not asking directly "Can you almost sense the smell of tobacco?"; instead, we state that she can almost sense it, and simply ask for confirmation ("can't you?").
- The belonging to the "most sensitive" category of people defined earlier ("*The most sensitive persons* almost feel the full taste of tobacco"): a negative answer would presuppose that she isn't concentrating hard enough, and that she doesn't belong to this class.
- The time left to the spectator to feel the suggested sensations.
- The friendliness and supportiveness shown so far: we are simply helping her to better visualize the scene.

When she confirms feeling this "almost sensation", the audience may also assume that she experienced everything else linked to it: feeling the full taste of the cigarette and the harder texture of its filter, seeing the gray smoke etc.

Out

What if she replies "No"? It is extremely rare to obtain a negative answer if you did everything as instructed, but people are strange. This line will take care of it:

Good, this means you're not so addicted yet!

As famed hypnotherapist Milton Erickson often advised, we graciously accept what we get by instantly redefining the answer's meaning in the most profitable way. This is a great out, with an embedded positive suggestion: she can still recover from her addiction, it's not too late *yet.*

If you resort to this out, it is preferable to skip the following "coughing" sensation and go straight for the "vibrations" suggestion.

IV.20   (Cough)

For now, let us continue with the description of the routine by following the most likely scenario, i.e. the spectator answered positively to the previous question.

Good, now many also mentioned a hidden wish to cough or [mime, clear the throat] *CLEAR THE THROAT* [gulp, silence]... *Can you notice something like that as well?*

The participant hasn't been speaking much in the past minutes, due to our long stacked patter and the relevant silences, and may in fact feel the need of clearing her voice before answering. In any case, she will reply positively for the following reasons:

- The fact of clearing *our* throat and gulping is often enough to make her feel "something like that"... When you hear somebody clearing his throat, you often feel an involuntary sensation to do likewise, or at least to swallow your saliva. It's a contagious cue, just like yawning.
- Coughing is "something like" clearing the throat. We once again leave her an illusory choice between two similar, linked options: an affirmative answer (she actually noticed a throat-clearing sensation) may have the rest of the audience assume that she actually felt like coughing.
- We are not talking about a plain sensation or feeling here, but a "hidden wish"... These coughing or throat-clearing wishes could be present, but she may not be aware of them yet, due to their hidden nature.
- Emotional people sometimes clear their throat when feeling uncomfortable or under scrutiny. Or, the participant may really have a smoking-related throat issue!
- She still wants to be part of the most responsive and sensitive category of persons (many of them also mentioned this hidden wish).

Out

In the unlikely case of a negative answer, you could resort to the previous out ("Good, this means you are not so addicted yet!"). This excuse will never be used twice, because if you had obtained a negative reply to the "Almost sixth sense" question (and consequently resorted to this out), you wouldn't have performed the coughing suggestion.

IV.21 Slow center

Good, now you can s l o w l y bring that cigarette closer to your right hand [pinch grip, mime] *and focus[11] on the center of your palm...* [Silence]

The *center* often has a prominent place in the sub-scripts: it suggests stability, control and focus. It is mentioned here for several other reasons:

- It is easier to concentrate on a specific point (in this case, the center of the palm) than on a larger, generic area.
- The center of the palm is a particularly sensitive area, which can easily amplify a number of unusual sensations.
- As we will see later, the tiny foil ball will be dropped in her slightly cupped hand and settle at its center.

When showing her how to bring the cigarette closer, use the pinch grip described earlier, pretending to hold it between your right thumb and first finger. The imaginary cigarette should be held almost in a horizontal position and point toward the center of the palm, as if preparing to enter the gap below the pack (Figure 10, shown with a real cigarette). Don't worry about the pack apparently blocking the way: all the focus is on the spectator's hand, not on the pack. The sensation she feels is what truly matters – especially to her!

---

[11] If the spectator missed the previous "concentrate" cue (during the induction), you may now want to substitute the word "focus" with "concentrate" and repeat the relaxation gesture / closed eyes association.

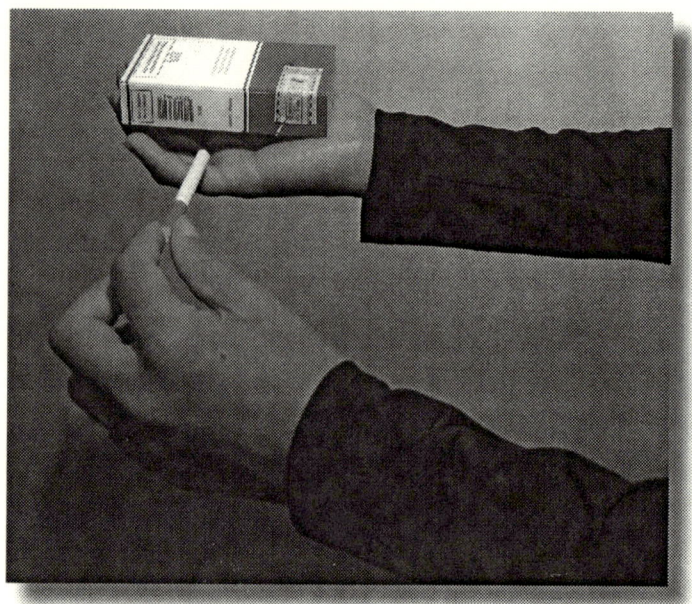

Figure 10

We are moving slowly and conditioning her to do likewise ("You can *slowly* bring that[12] cigarette closer to your right hand") to leave her more time to visualize the scene and feel the suggested sensations. Besides, these slow motions will be useful in the "Burning now" paragraph, where she will apparently decide the exact moment of the burn.

IV.22 Vibrations

And as you bring it closer [pinch grip, mime], *you should feel a slight warming vibration or something like a TINGLING sensation in your palm...* [Silence]... *You probably begin to be aware of it, don't you?* [Nod]

---

[12] Again, we are referring to "*that* cigarette" (over there), not *this* one here.

Let us analyze why she will answer "Yes" about 9 times out of 10:

- She has been holding her hand still, in mid air, for a while now: she will really feel a slight vibration/tingling. You should try this on yourself first, to realize that it happens and that you are not making anything up – you will feel much more confident during the performance.
- We have linked the sensations in the same sentence ("You should feel a slight warming vibration or something like a tingling sensation"). She may not feel warming vibrations, but quite certainly something similar to a tingling sensation; when she agrees, the rest of the audience will not know what she referred to and may assume that she also felt some kind of warming vibrations. By the way, it's a *tingling*, which sounds more generic than a *tinkling*.
- Of course we do not ask her brutally if she feels this sensation: instead, she may *begin to be aware* of it (the feeling could be present but she may not have noticed it yet).
- Just as in the "Almost sixth sense" paragraph, the last sentence is an affirmation coupled with a negative question ("You probably begin to be aware of it, *don't you?*"): we simply ask for an easy confirmation.
- We use a number of sneaky qualifiers: she "probably" begins to be aware of something, the sensation is "slight", it's "something like" that (a feeling similar to a vibration or a tingling – which are already unspecific). The range of sensations qualifying for our request is quite broad, isn't it?
- The conclusive nod is a physical trick exploited by clever salesmen who use their head: subtly nodding in an affirmative way at the end of a question helps obtaining a positive answer. Do not overuse it!

- Once again, we leave her all the time and silence she needs to experience the requested sensations.
- The constant supportiveness shown so far continues to play in our favor: she is now part of the most responsive and sensitive category of people, and has no reason to go against the flow and disappoint our expectations.
- The number of "Yes" previously obtained, and the fact that everything we said was true, also contribute in obtaining another positive reply.

Out

What if she answers "No"? Better prepared than sorry, you could use:

> *Not yet, that's okay, but please continue focusing on the center of your palm, because the more analytic people don't feel anything for a long time, then a BURNING* [burning gesture]... *the more sensible persons tend instead to notice an unusual sensation grow slowly as they relax even more...* [Silence]

Her "No" is immediately redefined as a "Not yet": remember, we said that she would *begin to be aware of* this sensation, so this relabeling is perfectly acceptable. Since she didn't notice it yet, we have to wait a little bit more in expectant silence. The double bind is quite explicit: she will feel something, period. It's just a matter of *how* the sensation will be perceived according to her personality and *how long* it will take for her to notice it, not *if* it will be felt. This out also covers the different outcomes of the routine: the sudden, sharp burning sensation caused by the gimmick, and the slow, gradual warming corresponding to the hypnotic suggestions. Besides, we now accept any unusual sensation ("Notice *an unusual sensation* grow slowly"), without even specifying its type.

IV.23 Warming-up

Very good, now if you continue bringing that burning cigarette s l o w l y closer [pinch grip, mime], *very soon you should experience a subtle increase in temperature...* [Silence]... *Can you notice your hand starting to get like a little warmer?*

She will reply affirmatively in most cases, for the reasons explained below:

- The increase in temperature is real: she has been standing with her arm extended for a while, with the pack resting on her hand (the palm may even be sweating a bit) and everybody has been staring at her expectantly all the time. In fact, her whole body may be getting a little warmer... Once again, we are not making anything up – simply try it on yourself! We now refer to her hand instead of the center of her palm, because the sensation is probably felt in a more general area.
- In *Psychokinetic Silverware*, Banachek reminded us that the item is not getting hot, but warm. In fact, not warm, but *warmer*. Actually, *a little* warmer. Or if you wish, *like* a little warmer.
- Many of the subtleties described earlier continue to be applied: the warming sensation is "starting" (which means that she may not be aware of it yet), the increase in temperature is qualified as "subtle", we

leave her enough time to experience it, everything we said so far was true etc. Please refer to the explanations in the previous paragraphs and make sure to individuate the relevant tools.

You may have noticed the continuous use of time words such as begin, become, increase, get, start, continue, warmER, closER etc.: the whole routine must be perceived (and remembered) as a successful process performed away from the participant and without touching anything. This will further minimize the importance of our tricky move in the chemical out/kicker.

Out

If the spectator hesitates for too long, or if you can sense that she is about to answer negatively, interrupt her flow of thoughts with:

Not hot yet, just a little warmer? [Silence]

We basically repeat the end of the previous sentence, highlighting the qualifiers: we acknowledge her difficulty in perceiving the change in temperature, and try to help her, making sure she understands the type of sensation to be felt. All we ask for is a slight and justifiable warmth.

This pattern interruption usually solves the participant's doubts. However, in case of a negative answer, strike back with:

Not yet, okay, but you really need to relax and amplify the unusual sensations you'll feel because most people notice them by now... [Silence]

Again, the time words used in our previous question allow us to relabel the "No" as a "Not yet". Any unusual sensation, felt anywhere, will do - and she even has to amplify it. This line must be said with a slightly resented tone: everything is in her hands, after all!

A negative reply here indicates that the spectator is not responsive: you may have selected the wrong subject, or you simply messed up with the script. It would then be wise to skip the next paragraph and go straight for the "Old Cowboy's last trick", which uses the chemical gimmick.

IV.24 Cause and effect: *Così è se vi pare*

The routine is becoming a bit long for close-up settings. However, advanced users with enough time in their hands may wish to continue the effect with their favorite hypnotic patters, if the previous responses of the participant were appropriate.

You could for example redefine any imperceptible, unconscious action you observed during the performance (but that the rest of the audience was unlikely to notice), as a consequence of her smoke addiction. For example, if you see that she is tapping her foot, one possible interpretation could be:

> *And when bringing that cigarette even closer to your skin, you would suddenly notice other signs of nervousness such as tapping the foot involuntarily and irregularly, meaning that the stress is increasing and the heat is on now...*

Once we saw the spectator tapping her foot, we defined the meaning of this involuntarily gesture in relation to the hypnotic experiment at hand, by associating it with the nervousness caused by the proximity of a lighted cigarette to her skin. In other words, we have reversed the cause/effect (or symptom/illness) link: we noticed an effect (tapping the foot), created a plausible meaning for it (nervousness, tension, stress) and finally made up its cause (proximity of a lighted cigarette).

On "Tapping the foot involuntarily...", the spectator will realize to her surprise that she was indeed doing it unconsciously, and stop for a brief moment (or at least slow down the tapping pace): we anticipate her reaction by adding "... and irregularly".

Here are a few other interpretations of the participant's possible actions:

- Scratching her skin: itching caused by smoking-related dermatological problems, or by previous cigarette burns.
- Sniffing: a kind of tingling in her nose, caused by the smoke of the invisible cigarette.
- Clearing her throat: we described this one already, and you will realize its full potential when she will actually clear her irritated throat during the experiment...
- Coughing: as above - caused by smoke-damaged lungs.
- Sweating or blushing: an increase in temperature caused by the proximity of the lighted cigarette to her skin.
- Biting her lips: dermatological irritation caused by the constant contact with the cigarettes.
- Tapping her foot or twitching her fingers: caused by nervousness, stress and tension - the typical nicotine withdrawal symptoms.

Remember to comment on those involuntary actions while standing at an angle and without looking directly at the participant, so it will seem that you couldn't have noticed them. Being slightly turned away is a great disarming technique: she will tend to relax even more and consequently allow those involuntary behaviors to surface.

IV.25   Breakpoint

When you feel that both the spectator and the audience are satisfied, you could interrupt the experiment with:

It's best to STOP NOW, before...

Let them mentally fill the blank, and compare this "best" option with anything that may happen otherwise. The "stop now" can once again refer to the experiment at hand or to stopping smoking in general. You would then skip the next paragraphs and continue with the reorientation patter described later.

For now, let us assume that you want to proceed with the powerful "burning" climax; I do not recommend using pure suggestions there, as the results obtainable with a gimmick are certainly more spectacular.

IV.26   The Old Cowboy's last trick

This is our chemical out/kicker. The move is named after the cowboy pictured on the ads of that famous red and white cigarettes brand: he died of a lung cancer a few years ago and became a no-smoking icon.

The beginning of our next sentence depends on the previous responses of the participant:

Case 1: you successfully obtained a number of unequivocal "YES". The sentence will begin with:

Great, and if you wish to continue even further...

We seem to give her the chance to stop the experiment, but we do not ask for confirmation and immediately proceed with the rest of the sentence (described later) to avoid interruptions. In any case, she probably doesn't mind continuing even further, because nothing bad has happened so far, she succeeded brilliantly and you congratulated her at every step. Please note that we remain unspecific about what is "further", because the gimmick could fail and the spectator may not drop the pack[13].

Case 2: you obtained insecure or forced "Yes", or an occasional "No". The spectator did not respond properly, and we offer her a way out:

Not yet, well to make it even easier and further speed up the process...

We are here to help and want her to be successful. Since she was a bit hesitant during the previous phases (or simply not responsive), we friendlily offer some assistance to make it even easier. Of course, never talk directly about "assistance" or "help", because it might play on her ego; don't even say "to make it even easier *for you*".

---

[13] This case will be described in the "Final out / Reorientation" paragraph.

Please note once again the reiterated use of time words such as yet, further, speed up, process. This experiment is a slow procedure (as repeatedly mentioned), and it takes time to reach the expected results: since the audience may not be willing to wait the whole day for satisfactory outcomes, we suggest speeding up the process to quickly reach its conclusive stage.

You should have completed the rubbing/folding actions in your left trousers pocket by now. Conclude the sentence with:

> ... *could you please read out* [step forward] *LOUD AND CLEAR that key sentence* [take pack], *SMOKING CAN...* [indicate "... SERIOUSLY HARM YOU", reposition pack]

The explicit warning on the pack will supposedly trigger some kind of subliminal response: since written and visual suggestions are notably stronger than verbal ones (depending on the type of audience, you may want to point this out before reaching for the pack), they will supposedly elicit a stronger reaction in the participant and help her feel the forthcoming burning sensation. Sure, smokers are already aware of the discouraging warnings printed on the packs in bold capital letters, but they choose to ignore them. We consequently remind the spectator of their existence by making her read the key sentence loudly, and by repositioning the pack with the writing in full view. In reality, the reason for touching the pack is to drop the foil underneath... This misdirecting excuse plays very, very well: the brief and justified repositioning of the pack is never questioned (and hardly remembered), and the simultaneous slipping of the foil passes completely unnoticed.

Before analyzing the move, please remember that this crucial moment should seem a natural extension of the successful experiment achieved so far, with the performer away from the participant at all times. (Almost) everything we said was true, and (almost) everything we predicted has happened; to walk one extra mile (i.e. to continue even further, or to make it even easier and speed up the process), we verbally highlight that trigger sentence. Other common notices are "Smoking is harmful to your health", "Smoking can lead to a slow and painful death", "Smoking can cause skin cancer", "Smokers are more likely to die young" etc. Every country has its own regulations, and imposes specific "catchphrases". Cigarettes producers are sometimes forced to print gory photographs on the packs, showing rotten lungs, burnt skin tissues etc[14]. If using this type of pack, make sure to point out that one picture is worth a thousand lives!

Now, the details of the move: slowly pick up the pack at the fingertips of your right hand, arm outstretched, without hiding it from the audience view (Figure 11).

Your left hand, secretly holding the foil in the usual pinch grip, indicates the writing with the index finger (Figure 12). For added misdirection, the right hand always moves earlier and faster than the left.

---

[14] See the recent European Union's anti-smoking campaign, discussed in the "Audience Assumptions" chapter.

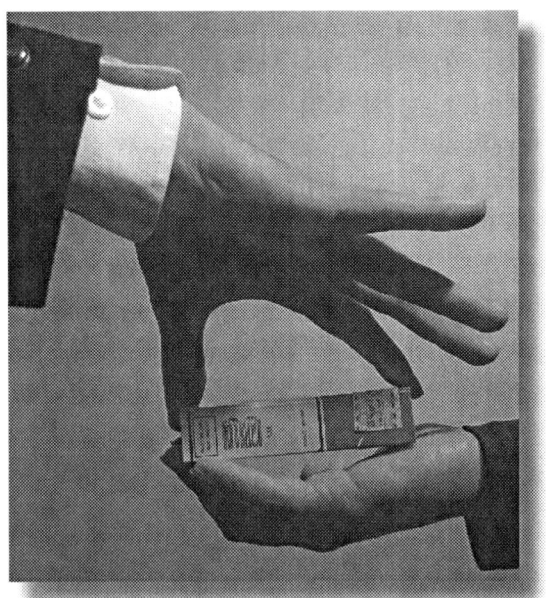

Figure 11

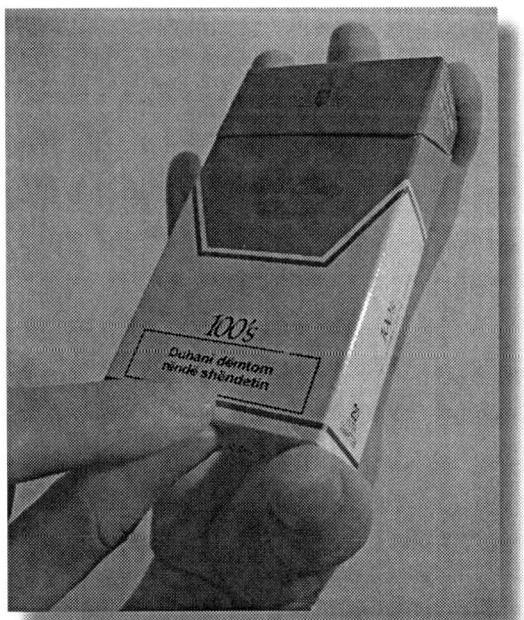

Figure 12

Replace the pack on her palm when she is reading the last words of the trigger sentence: your left hand will help the right in repositioning it, by putting the left index finger on its bottom and the thumb on top (Figure 13). The foil is pressed on the bottom of the pack by the left index finger, hidden from view. When replacing the pack in her right hand, you will naturally remove your left index finger from the bottom of the pack, allowing the tiny piece of foil to fall on her palm. Slightly step away in the described clockwise direction (audience view), and immediately continue with the "burning" patter explained later. Please note that your hands can be seen empty at all times, fingers apart, except for the fleeting moment when the foil is hidden in pinch grip.

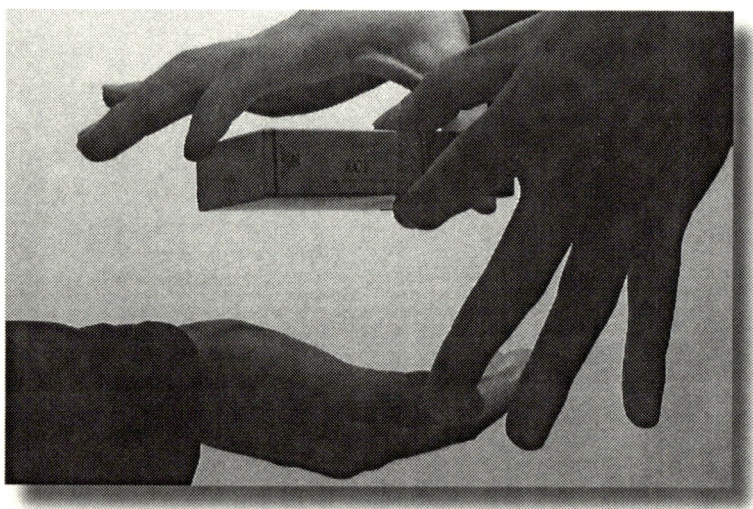

Figure 13

In addition to having the key sentence read out loudly, the pack is replaced with the writing side up, clearly showing toward the participant - that is, not only on the correct side but also with the proper alignment (Figure 14, spectator view). This repositioning non-verbally provides an additional justification to our brief handling of the pack (and, more generally speaking, to its presence in the routine). If by chance both the side and the alignment were already optimal when she originally placed the pack on her palm, you would of course replace it in the same position. I initially thought of taking advantage of this situation by pointing out how she had subconsciously placed the written/visual warning in full view (indicating for example that she instinctively realizes the damages caused by tobacco), but this would compromise our excuse for touching the pack. Mentioning it *after* having picked up the pack wouldn't be advisable either, for a question of time: due to the nature of the gimmick, the aluminum will reach its maximum temperature in less than 15 seconds.

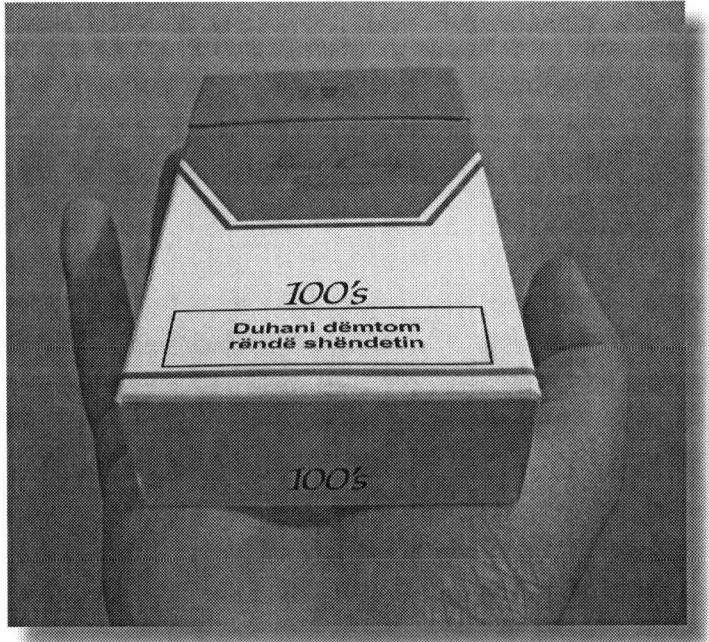

Figure 14

The piece of foil will touch her palm a fraction of a second earlier than the pack, but she will never be aware of it because:

- It is extremely small and light.
- It is falling from a very low height (about one half inch).
- The larger and almost simultaneous brushing of the pack on her palm will naturally cover any minimal impact below it (and before it).
- She is still focusing on the writing during the drop.
- You have shown your hands empty all along, and continue to do likewise after replacing the pack.
- She does not expect anything tricky since the experiment has been successful without you touching anything (even if you obtained an occasional negative reply among the many "Yes"). In fact, there doesn't seem to be enough time to perform any covert move: the pack is in your hands for less than 5 seconds, from "Smoking can..." to the completion of the sentence.

After the drop, don't worry about her feeling something strange in her hand: the foil is noticeable only when you know it's there. In any case, this subtle sensation could be attributed to a continuation/increase of the previous warming vibration or tingling noticed in her palm during the "Vibrations" phase.

Remember, no magician's guilt: avoid sharp movements and **never touch the spectator**. Practice this move until it is second nature – it's the only one in the entire routine.

IV.27   Additional misdirection

When you pick up the pack, the spectator may inadvertently gesture with her left hand (which supposedly holds the invisible cigarette). In this case, I discreetly slip in the following line:

Be careful with that cigarette!

We verbally minimize this command by slipping it in the middle of the previous sentence, with a lower tonality and a faster pace: in his *Wonder Words* audiotapes, Kenton Knepper called this technique "Dropping the bomb" (one couldn't find a more appropriate name for it). It further distracts her from the drop of the foil in her right hand, and hides a nice subtext. I am sure you would never say something like "KEEP THIS CIGARETTE here", which would contradict all previous suggestions.

IV.28   Alternative handling

I initially played with a number of different handlings, but the "trigger sentence" justification is the best I could find. Just for the records, one of the excuses I was using to drop the foil was to move the pack from the participant's right hand to the left, stating:

That's because you don't want to injure your good hand in case something goes wrong!

If for some reason you decide to use this line, please do not say "That's because *I* don't want to injure your good hand", but "That's because *you* don't want to injure your good hand": it's her choice, and everything is in her hands: *she* doesn't want to injure herself.

IV.29 Burning now

It would be even more convincing to have the spectator repeat the key sentence several times, but we are running against the clock: the hardest part of the routine is not the "move" (which is in itself very simple), but the necessity to keep talking, gesturing and breathing calmly afterward, without changing by any mean your pace and tonality. This is not so easy when you know that a time bomb is about to explode... The forthcoming climax, i.e. the spectator dropping the pack, must appear as a natural extension of the effect: as mentioned earlier, you are simply continuing a successful process, and the experiment's conditions must seem unchanged.

As soon as she finishes reading the sentence, proceed with:

> *Exactly,* so please Sue continue relaxing [hand positioning] *and WHENEV...* [clear the throat] *WHENEVER YOU WANT, you should s l o w l y bring that cigarette even closer and carefully STUBBB IT OUT* [pinch grip, mime] *on your hand* [step backward, silence]

To further confirm that everything is under her control, she will now apparently decide the precise moment of the burn. The words "Whenever you want" highlight her freedom of choice, and are marked by repetition, fumbling ("Whenev... whenever you want"), throat clearing (which is also related to the previous "coughing" suggestion), louder tonality and direct eye contact with the participant.

The audience would become suspicious if the burn happened immediately after the handling of the pack; however, the time misdirection offered by the gimmick is pure gold as it distances the climax from the move, canceling any thought of trickery.

Now for the bad news: it is almost impossible to accurately predict the heating moment, since it depends on numerous factors:

- The conditions of the pellet.
- The typology and size of the foil.
- The amount of chemical on your fingers.
- The pressure applied when rubbing and folding the foil.
- The sensibility of the participant.
- The position of the foil in her palm.
- The gap between the foil and the pack.
- The general weather conditions (temperature, humidity).

There are too many variables to achieve a perfect timing. This is why we conditioned the spectator to act slowly since the beginning of the effect, by repeating this word many times and by moving, gesturing, speaking and breathing accordingly. We sympathize and understand her hesitation in stubbing out that cigarette on her hand, since all the previous sensations were felt just like if a *real* lighted cigarette had been used: she certainly needs to do it with great care... In reality, this cautious slow motion will be employed to make sure that her "decision" coincides with the chemical reaction.

Before considering the different scenarios, let us highlight a few additional subtleties in the script:

- If you noticed her right fingers straightened and tense after replacing the pack in her hand, you would repeat the "curling fingers" visual cue: once again, establish eye contact when calling her name, drag her look toward your right hand and curl your fingers on "relaxing" ("please Sue continue *relaxing*"), with your hand at stomach level and close to your body.

- We have marked the pronunciation of the keyword "stub" ("carefully *STUBBB* IT OUT on your hand") to have it easily recalled later: we will return to this issue in the misremembering section.
- After miming the stubbing action, step backward as to further distance yourself from an imminent danger, giving the audience a clear view of the climax: you are now completely out of the picture, just as in the beginning of the routine (but this time, on the left side of the participant).

IV.30   Multiple outs and nominalizations

There are three possible outcomes:

Case 1: the chemical reaction takes place exactly when the spectator decides to stub out her cigarette. Of course, this is the perfect situation: the foil becomes hot when she pretends to burn herself. No adjustment would be required in this unlikely scenario.

Case 2: the chemical reaction takes place *before* the spectator's decision. In other words, she feels the heat before thinking of stubbing out her cigarette. Believe it or not, this is not a problem. The imaginary cigarette was already near her hand at the beginning of this phase, and she was bringing it even closer: nobody knows for sure the length of that invisible cigarette, probably the lighted end was already close enough to burn her, even if she didn't think so... The rest of the audience will never be aware of this delay, and the spectator herself will never mention it: even if the moments seem to differ, she is so overwhelmed by the burning sensation that this slight and justifiable inaccuracy will never be questioned. You will find out when performing the effect that there is no need to point out an excuse - the participant will mentally make it up by herself.

Case 3: the chemical reaction takes place *after* the spectator's decision. In other words, she pretends to push the imaginary cigarette on her palm before feeling the burning sensation: she brings the invisible cigarette closer and closer, while everybody is staring at her expectantly, and nothing happens... To avoid this embarrassing situation, we will once again rely on a pattern interruption. If she ever tries to speak, or if you see that her left fingers are getting too close to her right hand, interrupt her immediately with:

> *Please YOU NEED TO STAY focused and be careful now...* [Silence]

This will take care of it: she needs to focus more intensely and carefully stub out that dangerous cigarette. It can be a bit nerve-racking but there is no need to worry, since the foil will eventually become hot.

To avoid objections, the commands are presented as necessary ("you *need* to stay focused") and linked ("stay focused *and* be careful"); asking two things at the same time generally reduces the risk of protests, because people have to think about which action they are going to question. The embedded command "you need to stay" may also suggest to stand still.

Finally, let us highlight one last tool used to get out of troubles. Nominalizations (the infamous "politicians' words") are extremely valuable in outs: words such as relax, concentrate, focus, etc. do not have a specific, tangible meaning (as opposed to "scar", for example), and the spectator needs to focus inward in order to assess what they mean for her. In our case, we asked her to "stay focused" and "be careful": those requests do not mean much when you think about it, but they do the job nicely since unspecific orders are more difficult to challenge. You may want to reread the previous paragraphs and individuate all the nominalizations... This technique is involuntarily used in everyday life as well, when people can't (or don't want to) tackle a specific problem. They may resort to words such as belief, trust, concern, trouble, depression, sadness etc. to describe an issue without specifically addressing it.

When the foil finally gets hot, the delay initially perceived by the participant will be intuitively justified by the unknown length of the invisible cigarette (in this case, it may have been shorter than expected) and by her initial lack of focus. And once again, she will be so surprised by the burn that the slight discrepancy between the moment she first *thought* of stubbing out her cigarette and the instant she felt the pain becomes irrelevant. Practice will give you the correct timing to reduce this delay as much as possible.

In conclusion, the audience will assume that the burning sensation felt by the participant was a direct, simultaneous consequence of her decision to stub out the imaginary cigarette on her hand; the participant herself will never question that.

IV.31   Climax and anchoring

During the chemical reaction, she will hastily reverse her hand (with a scream, if you chose the right subject) and consequently drop the pack. The foil will fall unseen by the audience for the following reasons:

- Its small size.
- The simultaneous fall of the pack (and of a few cigarettes, if she left it opened when initially pretending to extract the invisible cigarette).
- The distance from the participant. If you film the effect, you will notice how the falling foil is virtually invisible even when you know it's there.
- Her surprised reaction and facial expressions, which will catch the audience attention at first.

The participant herself won't notice the foil, since she will hurriedly drop the cigarettes and promptly check her empty palm to make sure that everything is all right. The falling pack is not the first priority: her graceful, delicate hand is. There won't be any trace left there either: the maniacal purists will appreciate the fact that the spectator usually rubs and massages her palm before even looking at it, removing any possible microscopic residual dust.

Immediately after the climax, exclaim:

EXXXCELLENT!? [Clap hands, silence]

This is the anchor I use during most climaxes - feel free to shout your favorite enthusiastic exclamation. For those unfamiliar with this technique, an anchor occurs when a subject is at the peak of an intense state, which is neurologically linked to a specific stimulus: in theory, this state could be repeatedly reached by setting off the stimulus[15]. Anchoring is particularly useful when you plan to involve the participant in a dual-reality effect later in the show: the ambiguous climax will generally require her to show a greater reaction than what she would normally express, to make it consistent with the audience surprise and expectations.

Our anchor is marked by:

- The exaggerated pronunciation of "EXXXCELLENT", which will make it easily recognizable later.
- The loud, enthusiastic and surprised tonality, in contrast with the softer and deeper tone used so far.
- The simultaneous, noisy clapping of your hands (which is sometimes interpreted as an applause cue by the audience...).
- You may also decide to utilize her name ("EXXXCELLENT, SUUUE!?"), but I prefer to preserve this keyword for establishing eye contact, as previously explained.

---

[15] Anchoring was introduced by Russian physiologist Ivan Pavlov in the end of the nineteenth century. In one of his famous experiments, he rang a bell every time he showed a piece of meat to a dog: after several times, the sound of the bell was enough to make the dog salivate, independently from the actual presence of the meat. Pavlov used the results of these experiments to develop his stimulus/response theory (*The Experimental Psychology and Psychopathology of Animals*).

IV.32 Decisions

After a few seconds of respectful silence, continue with:

Right decision, what did you feel exactly?!

Her "right decision" can refer to the fact of having stubbed out her cigarette, or to the *moment* when she decided to do it. The open-ended question "What did you feel exactly?!" further distracts her from the possible timing discrepancy, and makes her focus on the *exact* type of sensation instead. It also allows for a number of interesting answers: pay attention to what she says, not just for a question of basic courtesy, but because you should memorize a few keywords to be repeated in the forthcoming misremembering patter.

I usually act very confidently when the effect is risky, and feign surprise when it is fail-safe. It should seem that the fall of the pack was beyond your wildest expectations: you were hoping for some kind of burning sensation, true, but certainly nothing of that intensity! The spectator over-delivered, she was the most responsive and sensible participant you ever met... Listen to her comments and leave her the scene: as part of a unique experiment, she will probably exaggerate the description of the effect when telling the story to her friends.

Before continuing with the analysis of the script, let us examine a rare situation, where the spectator did NOT drop the pack: the foil did not get sufficiently hot, and she is still standing there with the cigarettes (and the foil) in her hand...

IV.33   Final out / Reorientation

When ideas fail,
words come in very handy.
Johann Wolfgang von Goethe

Or in our case, language replaces faulty gimmicks.

From the audience point of view, the experiment has been successfully completed: the participant felt a heating sensation when pretending to stub out her cigarette (and she certainly acknowledged it) – nobody ever promised that she would have screamed or dropped the pack. The problem is that you don't end up clean: she will now want to check her palm, by raising the pack and... try to explain that little baby underneath. I don't know about you, but I masochistically cherish those moments. The piece of foil laying under the pack would not explain how the experiment was achieved, but it would certainly suggest that something tricky occurred.

Before analyzing the out, let us list the top ten causes of this problem, in order to prevent it:

- The pellet was old or used-up.
- You had filthy pockets or dirty hands.
- You were performing in cold and dry settings, and didn't moisten your fingers before the setup.
- You did not apply enough chemical on your fingers, or you set up too early and touched something before the performance.

- The piece of foil was too small, or you didn't rub it properly.
- You did not use pure aluminum foil, and rubbed the paper side.
- The spectator did not curl her fingers and the pack completely covered the foil, hampering the heating process.
- You overlooked the previous conditioning and/or seeding suggestions.
- You chose a macho spectator who stoically resisted the temptation to drop the pack.
- You rubbed a breath mint and ate the pellet instead.

I am naturally speaking from experience here (except for the last point). While all these parameters seem under you control, mistakes must be taken into account. This is how you can deal with it:

Great, now please Sue you really need to CONCENTRATE [downward gesture, close eyes] *a little more and take a deep breath* [inhale] *as you feel the renewed energy in your body, and shake your han...* [remove pack] *- KEEP CONCENTRATING and shake your hands to reactivate the blood flow as you take another deep breath and focus on the voices of your friends, returning safely from your mental journey now* [upward gesture]

The excuse to get rid of the foil is to have her shake her hands to "reactivate the blood flow", since she has been holding them in mid-air for a while... We consequently lift the pack for a moment so that she can shake them freely (we obviously don't want her cigarettes to fall on the floor!). To make her close her eyes, we use once again the "concentrate" trick: establish eye contact when calling her name and close your eyes on "concentrate" ("Great, now please Sue you really need to *concentrate*"), while doing the "downward gesture" explained earlier. Don't move nearer until removing the pack, because she will feel safer in closing her eyes if you stand away from her. In case of a non-responsive spectator, you would of course directly ask her to close her eyes, to better focus on the renewed energy in her body. "Keep concentrating" is said right before lifting the pack, and means "Keep your eyes closed" to the participant. The rest of the audience, unaware of the concentration/closed eyes association, simply understands that she should continue focusing despite the brief disturbance you caused by removing the pack. She "really *needs*" to maintain this altered state until the end of the experiment – ever wonder what would happen otherwise? People know that interrupting a state of trance can lead to traumatic consequences...

To avoid interruptions, we used a stack of linked commands (concentrate / take deep breath / feel energy / keep concentration / shake hands / take deep breath / focus) without giving her the chance to speak or think about anything else: the effect is over, she was successful and has no reason to argue.

After lifting the pack at the fingertips of the left hand, keep it on top of hers (Figure 15) until she drops the foil: this precaution ensures that she won't see her right palm if she unexpectedly opens her eyes during the removal of the pack. It is wise to keep looking at her straight in the eyes to "hold" her look in case she opened them too early, and immediately remind her to "keep concentrating". In close-up settings, you should stay in front of the participant to conceal the foil from the audience view. Once she dropped the foil, promptly replace the pack on her palm, and continue with the script while slowly stepping backward in the described clockwise direction. I know, it would have been better not to touch anything, but the effect was over anyway (?)

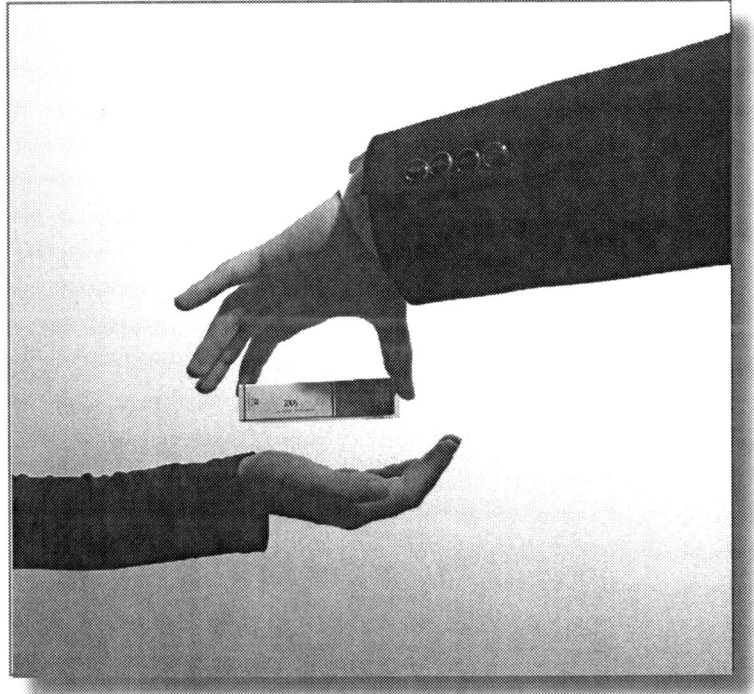

Figure 15

Don't underestimate this paragraph: even if you may never resort to the out, you could use this patter as a conclusive speech indicating the end of the effect – just like a reorientation process after the trance, or a relaxation moment following the difficult test that the spectator had to pass. Reorientation processes are usually accomplished by making the subject concentrate on sights and sounds in the immediate surroundings, such as the hypnotist's voice or the audience's noises (*"Focus on the voices of your friends*, returning safely from your mental journey"). The relevant upward gesture on "returning safely from your mental journey" is basically the downward gesture done in reverse, with the opposite effect and one common objective: indirectly showing your hands empty. Your pace and tonality have also gradually returned to normal, after the deeper hypnotic tone used during the experiment - a message to the subconscious mind of the spectator, suggesting a return to its normal waking state.

We have also slipped in the positive suggestion of a renewed energy in her body... a natural consequence of our successful experiment. As previously mentioned, stubbing out the cigarette and letting go of the pack are a metaphor for stopping smoking: it took some time and courage to "stup", it was a bit painful but certainly worth it. The spectator now notices this renewed energy, and feels much better.

IV.34   Better now

You feel better now, don't you?

She definitely does, because this scary experiment has been completed brilliantly, the burning sensations are over, and she checked that her hand was all right. Another "Yes" added to your conspicuous set. Of course the hidden message is, once again, that stopping smoking (i.e. stubbing out the cigarette and getting rid of the pack) makes her feel better.

IV.35   Post-hypnotic suggestions

Apart from the rare case described in the "Final out / Reorientation" paragraph, the spectator dropped her pack on the floor and will now want to retrieve it. I initially picked it up myself and gallantly returned it to its owner, before coming across a clever idea described by Kenton Knepper in *Kolossal Killer*: start reaching for the pack, then step back as if you realized that you shouldn't be touching anything, and let the participant retrieve it herself. This subtlety non-verbally reinforces the suggestion that you never did anything at all. At first, I wrongfully feared that the foil on the floor would be noticed (I even thought of stepping on it while pretending to pick up the pack); however, it is virtually invisible and does not require thick carpets or dirty restaurant floors. Spectators don't see it, or at least don't notice it: they are not remotely aware of the Hypnoheat gimmick, and can't link a tiny piece of paper that could have been laying on the floor for days to the remarkable hypnotic experiment they just witnessed. Besides, the participant is focusing on her pack, and is also busy listening to your first misremembering question.

IV.36   Thirteen Steps to Misremembering

It is now time to briefly summarize the effect. Most spectators wouldn't be interested in listening to the detailed description of the events that just occurred before their eyes, but they often appreciate a concise recap of the experiment (just like an instant replay on television). We obviously need to sum it up, by omitting irrelevant details and focusing on the alleged core of the routine. For example, a floating bill trick could be summarized as "A magician made a bill float in mid-air"; or, "A magician borrowed a bill, rubbed it on his trousers to create static electricity and made it float in mid-air"; or, "Earlier this afternoon, an elegant guy with a funny moustache and a lovely wife - I think his name was Mickey or something - asked for a bill: he first wanted 50 or 100 bucks but he couldn't obtain them, so my cousin Vinny, the one who lives close to that used snowmobiles dealer, was kind enough to lend him a dollar etc."

Start the following patter when the spectator is picking up the pack:

> *So, BASICALLY you stood over there and pretended to smoke an invisible cigarette* [pinch grip, mime], *yes? And you could almost sense* [inhale noisily] *like it was a real one, couldn't you? And as you were s l o w l y bringing that cigarette closer to the center of your right palm* [pinch grip, mime], *it became hotter, isn't it?* [Nod] *And when you finally decided to STUBBB IT OUT on your hand, you felt...* [use her own words, exaggerated burning gesture], *is that correct? You may not know exactly why you felt all those burning sensations and this compulsive urge to THROW THE CIGARETTES AWAY* [exaggerated burning gesture], *but you should try to imprint that moment on your memory for the rest of your life, and recall it every time you will try to smoke a cigarette, all right?*

Keeping in mind that memory can be creative, let us analyze in details the various misremembering tools applied - I have bought a pair of burgundy velvet gloves to type the following paragraphs (the clerk was looking at me strangely, but who cares).

1. Memory breakdown

Break up the summary into several short sentences, asking each time for confirmation. This seems a proof of absolute fairness, and gives the impression of seeking the approval of the spectator at each single step. In reality, you are the one selecting those steps, and consequently choosing which details should be omitted, which actions should be emphasized, and in which order they should be mentioned. This technique is often used to alter the order of the events: what you say is "basically" true if taken sentence by sentence, but the analysis of the patter as a whole might instead reveal your time alterations and faulty omissions.

Breaking up the summary into separate questions also allows you to obtain additional positive answers, which will rapidly add to your Yes Set and eventually outnumber the occasional "No" you may have received during the performance.

2. Aural assumptions

Make sure that the spectator audibly confirms each mis-remembering sentence, for everybody to hear. A passive nod of the head wouldn't be enough, because:

- The audience doesn't usually trust performers of your kind. A participant remaining silent wouldn't necessarily mean that you are speaking the truth: your affirmations are much more convincing when she audibly agrees.
- The participant may fool herself by confirming your statements.

- She is more involved, and will therefore (mis)remember the experiment more vividly, further exaggerating its description with her friends.
- As discussed above, this technique provides a number of additional "Yes" in quick succession.

3. Omissions

We are "basically" stating the truth, yet skipping over a few seemingly irrelevant details (just like in a deceptive magic ad). Some examples of remarkably unimpressive omissions:

- During the induction, we came within reach of the spectator to supposedly light her invisible cigarette (it was a gag of no importance).
- During the "Old Cowboy's last trick", we handled the pack (it was simply meant to highlight the trigger sentence).
- In the "Final out / Reorientation" paragraph, we lifted the pack for a moment to allow her to shake her hand (the effect was over anyway).

The participant will never question our summary, because those actions were psychologically invisible at first and do not seem significant enough to be mentioned.

4. Time alterations

This technique refers to the modification of the events' duration, and to the alteration of their chronological order. We considerably shortened a number of occurrences through omissions, and modified their lifespan: for example, we said that her hand got hotter ("As you were slowly bringing that cigarette closer to the center of your right palm, it became *hotter*"), while in fact it only became "a little warmer" at first, and then burned for a fraction of a second.

5. Exaggerations

In addition to the warmer/hotter upgrade described above, there are a number of emphasized statements:

- "You may not know exactly why you felt *all those burning sensations*": in reality, there was only one burning sensation (during the climax), but who knows for sure how many were noticed among the previous tingling, warming etc.? Besides, the question is not their number, but if she knows *exactly why* she felt them (presupposing that she actually felt them).
- "And this *compulsive urge* to throw the cigarettes away": her natural reaction to the burst of heat caused by the chemical has been relabeled as a flamboyant "compulsive urge".
- "*Throw* the cigarettes *away*": the spectator has violently "thrown away" the pack, she didn't just drop it with nonchalance. Or at least, that's what you want people to tell their friends. This wording sounds much more powerful, and is perfectly acceptable because the spectator usually overreacts during the climax: her surprised, uncontrolled movement could be defined as a "throw".
- "Throw *the cigarettes* away": "the cigarettes" refer to the pack and the invisible cigarette previously held in her left hand. There seems to be more than one thing, and it sounds more effective than "Throw *the pack* away". Once again, the point is not what she threw

away, but rather if she knows exactly why it happened. Besides, we do not ask for confirmation: the linked question refers to her will of trying to remember this event for the rest of her life.

- Concerning the visual exaggerations, the burning gesture itself is accentuated and mimed in full view this time, with your extended right arm.

6. Devil in the details (reprise)

Mentioning true details helps covering important facts that were strategically omitted, and adds credibility to your summary. For example, we said "As you were slowly bringing that cigarette closer to *the center of your right palm*": we could have simply referred to her hand, without pinpointing the center of the right palm, but the three details we added are true, harmless and make our summary appear more specific and complete than what it really is.

7. True lies

There is a thin red line between exaggerations and lies. What we say has to be basically true, at least when considered sentence by sentence. For example, we stated "You pretended to *smoke* an invisible cigarette", not "You pretended to *light* an invisible cigarette": the latter would be a plain lie, since you were actually the one pretending to light it, in order to move near the participant. I made this mistake several times, until a smart spectator said jokingly "No, YOU lighted it". It got a laugh (since we are talking about invisible items, after all) and the effect was concluded smoothly; however, you don't want to hear any "No" during hypnosis acts.

Of course, you could play with this idea by lying on purpose and promptly pointing out your own mistake, before linking the corrected sentence to an even stronger compromise...

8. Repetita iuvant

Repeating the exact same words used during the performance proves that we are faithful to the original facts, and ultimately means that we are loyally describing the same actions. Aren't we? This is in my opinion a crucial step in the misremembering process: a sequence of truthful events, described with the same words, helps eclipsing the intentional inaccuracies and omissions. For example:

- "You could *almost sense* like it was a real one, couldn't you?" is similar to "You can almost sense its smell, can't you?" (in the "Almost sixth sense" paragraph). The noisy inhalation cue on "almost sense" will remind her of the smell reference.
- "You were *slowly bringing that cigarette closer* to the center of your right palm" is similar to "You can slowly bring that cigarette closer to your right hand" (in the "Slow Center" paragraph), "If you continue bringing that burning cigarette slowly closer" ("Warming up") and "You should slowly bring that cigarette even closer" ("Burning now"). The word "s l o w l y" is also pronounced with the same unhurried pace of the previous occasions.
- "When you finally decided to *stub it out on your hand*" is similar to "Carefully stub it out on your hand" (in the "Burning now" paragraph). To further ease the identification of the keyword "stub" (an unusual word in itself), we had originally exaggerated its

pronunciation ("STUBBB"), and obviously repeat it now with the same emphasis.

9. Self-referencing

To take the previous concept one step further, we also repeat the exact same words used by *the participant* when describing her sensations (in the "Decisions" paragraph): those words are certainly true for her[16]... You would of course repeat easily-remembered keywords – including possible swearwords pronounced by the spectator, if you are a Bob Cassidy fan. For example, if she answered "I felt something like a sudden sharp sting in my hand", the important keywords would be "sudden", "sharp" and "sting". The relevant misremembering sentence would sound like "When you finally decided to stub it out on your hand, you felt something like a sudden sharp sting, is that correct?"

Incidentally, this shows your alertness and carefulness about her feelings.

10. Presuppositions

This tool has already been employed in the suggestions phase, but this time we apply the self-referencing method described above: we link our own, arguable statement to what the participant herself previously mentioned – something she simply can't refute. The question is not necessarily related

---

[16] In an interesting article published in the February 1999 issue of *Legal and Criminological Psychology*, Giuliana Mazzoni, Manila Vannucci and Elizabeth Loftus noted that "Self-generation and self-referencing are factors that modify the memory quality of false recognitions".

to the previous affirmation, but when the spectator answers positively, the audience will assume that the linked sentence was true as well. This technique is extremely powerful in the misremembering phase – you can get away with the most daring presuppositions! To continue with our example, the presupposition "When you finally decided to stub it out on your hand" is not entirely true, since most of the time the moment she took her decision slightly differs from the instant she felt the pain; the linked question however ("You felt something like a sudden sharp sting, is that correct?") is related to something she said herself a few moments ago. This will invariably bring a positive answer, and the audience will take the linked presupposition for granted.

11. Choices

The participant has taken a *decision* ("When you finally *decided* to stub it out on your hand"). The conclusive speech is the best moment to remind it, even if the choices were forced or strongly compromised. The most common example would be a spectator "selecting" a forced card: she is first asked to simply "take" a card, and she will later be told that she has "chosen" it. In our case, the participant didn't select much, but the rest of the audience will remain completely unaware of this inaccuracy – they may in fact remember that we clearly stated "Whenever you want…" when asking her to stub out that cigarette.

12. Symbolisms and metaphors

The conclusive speech offers additional subtexting op-portunities and can reinforce previous suggestions, meta-phors and symbolisms that were not perceived in the "heat" of the moment (you are right, it is best to STOP now).

For example:

- Stubbing out her cigarette ("When you finally decided to stub it out on your hand") and dropping the pack ("This compulsive urge to throw the cigarettes away") could represent, as previously explained, her decision of stopping smoking. Please note the embedded command "Throw the cigarettes away", marked by a louder tonality and direct eye contact with the participant. We also hinted at the overkill assonance between "stop" and "stub".
- "Every time you will *try* to smoke a cigarette" suggests that she may not actually succeed in smoking it, thus confirming the validity of our therapeutic experiment.

13. Forever

The best way to have a spectator (mis)remember something for the rest of her life... is to ask her to do so! Or better, ask her to *try* to do so ("You should *try to imprint that moment on your memory for the rest of your life*"): at this stage, she may not know for sure if she will remember it for the rest of her life, but she will most certainly try to do so.

If in a darker mood, you may also want to remind her that sooner or later, everyone quits smoking.

IV.37   Conclusion and covert applause

Thanks again, you really did GREAT.
CONGRATULATIONS!! [Applause cue, shake hands]

We acknowledge once again the crucial role played by the spectator in the experiment (say it like you mean it!); whatever really happened, *she* did great and *we* succeeded brilliantly. Unless performing in intimate close-up settings, the enthusiastic "Congratulations!!" and relevant applause cue should bring you a round of applause that you will graciously leave to the star of the show. Before the end of the clapping, thank her again while indicating her seat, and get ready for your next effect.

IV.38   Clean up

If you used a chemical gimmick, don't forget to discreetly clean up your left fingers on the fabric of your clothes; you will then thoroughly wash your hands at the end of the show by using a proper detergent. Please don't omit this last move, as it is an integral part of the routine.

V. A NOTE ON SCRIPTS

The memory work necessary to perfectly recite the patters of an entire show is quite hefty. It can be done of course, since you don't need to remember the motivations behind each word or gesture: while it took a long time to explain, the full script of this effect is only a few pages long – much less than what a theater actor memorizes in his daily job. However, in real close-up situations, prompters are not practical, spectators can interrupt you and the participant can react in many different ways: it is almost impossible to pull off all the subtleties without the slightest mistake. Or at least it is for me: when watching my performances on video, I always noticed some kind of imperfection in the tonality, timing, body positioning etc. My goal has always been to stick as much as possible to the most complete script I could devise, deviating only when my memory failed or when disturbed by unpredictable events. Despite those imperfections, the effect is in most cases completed successfully, even if somewhat underachieved.

You will of course need to modify the script, customizing it to your presentational preferences and performing style. Hopefully the details provided will allow you to do so without altering the crucial psychological ploys that bring the routine to its full potential: they make the difference between a good effect and a memorable (and inexplicable) experience.

VI. VARIATIONS

Let us briefly describe three variations of the effect: a non-smokers version using your own pack of cigarettes, a no-pack presentation and a beginner's version I performed many years ago. The scripts will be almost identical to the one described, with a few minor modifications: this will allow you to continue using most of the original patter, and save your memory for more important issues. I find it very convenient to learn a single, robust script, and change only a few words according to the situation at hand.

VI.1 Non-smoking audience

At the beginning of the routine, you enquired about the number of smokers in the audience ("How many smokers do we have here... could smokers please raise their hand for a second?") but found out that nobody smoked. Good for them. As previously mentioned, we will adjust the patter and present a prevention experiment instead of a curing therapy. Respond with:

Nobody, great, that's a health-conscious group!

Once again, we gladly accept what we get, and congratulate the audience even if they involuntarily boycotted our primary effect. No further justification needs to be provided at this stage, since the audience doesn't know what we intended to do in the first place.

After having selected a spectator meeting the described criteria, substitute the question "How did you start smoking?" with:

How come you don't smoke?

She may talk about health concerns, the constant hassle of secondhand smoke, friends asking her to try, the risk of feeling out of certain "cool" groups, the alternatives to tobacco, the usefulness of governmental anti-smoking campaigns etc.

A humorous possibility for teenage participants: after she finished explaining why she doesn't smoke, ask her to bring out her cigarettes, using the same line of the smokers' version ("So please bring out your cigarettes"). This request originally appeared legitimate, and the chuckles were due to the hesitation of the participant in giving out her precious cigarettes to somebody who wanted to make her stop smoking. In this case however, the sentence is ironical, as if we didn't believe a single word of what she just said and assumed that she was trying to hide her little smoking secret from her parents; once again, this will bring a few chuckles from the audience. On a theoretical level, the linguistics may be worth pondering: same words, same results but different interpretations.

You will then substitute the rapport-building sentences ("My father used to smoke too, but I slowly made him stop" and "I can show you how I did it, if you think you may try to stop too one day" in the "Rapport" and "Yes Set" paragraphs, respectively) with something relevant to her answer - possibly bringing the conversation toward the importance of preventing the addiction, instead of curing it.

In the "Framing" paragraph, you will of course substitute the word "cure" with "prevent" ("Let me show you this brief hypnotic experiment used to *prevent* smoke addiction"). If you are a smoker, one may wonder why this technique was not used on you, to prevent you from smoking: the excuse this time could be that you learnt this method too late.

You will be using your own cigarettes. However, please remember our introductory discussion about props: removing a pack of cigarettes from your pocket is acceptable when you are a smoker, or if the spectators ignore you don't smoke. Otherwise, it would create suspicion. Select a brand which is common in the country you are performing, to quickly establish its legitimacy. If possible, the pack should have some bright reddish colors (suggesting heat) and the written or visual warning should be easily linkable to the burning suggestion (generic harm, skin burns, damaged lungs etc.). Do not use a brand new pack, because the spectator would waste time in removing the plastic wrapper when supposedly extracting her invisible cigarette. Besides, this plastic seal suggests some kind of protective, incombustible shield around the pack.

The rest of the patter will remain basically the same because we never specified the owner of the pack (in fact we never mentioned the pack at all), but only referred to the invisible cigarette and the participant's hand. The script stays unchanged even in cases where a non-smoking environment might seem incongruent. For example, the smell/taste senses in the "Induction" and "Almost sixth sense" paragraphs can still be applicable to non-smokers: due to secondhand smoke, people know (or at least "almost sense") what a cigarette smells/tastes like even if they never smoked anything in their life[17]. You would of course put yourself on their side by highlighting the unpleasant stink of secondhand smoke.

People always say that
my smoking is bothering them...
Well, it's killing me!

Wendy Liebman

In the out of the "Almost sixth sense" paragraph ("You can almost sense its smell, can't you?") or of the subsequent coughing suggestion ("Many also mentioned a hidden wish to cough or clear the throat... can you notice something like that as well?"), you would substitute the excuse "Good, this means you're not so addicted yet" with:

Good, that's because you're not a smoker!

At the end of the routine, you will still pretend to retrieve the pack from the floor and step back without actually touching it, even if it's yours: let the spectator pick it up herself and examine it at her heart's content. The misremembering speech also

---

[17] For extra safety, you could substitute "The most sensitive persons almost feel the full taste of tobacco" with "The most sensitive persons almost feel the first lungful of tobacco", removing the taste reference.

cuts both ways: "Recall it every time you will try to smoke a cigarette" is valid both for smokers who may try to smoke again after our hypnotic cure, and for non-smokers who may try to have their first cigarette despite our preventive therapy.

VI.2   The cowboy is dead

When I don't smoke I scarcely feel as if I'm living.
And I don't feel as if I'm living unless I'm killing myself.

Russell Hoban

If you are performing for non-smokers and don't want to use your own cigarettes, you can deliver the prevention patter and ask the participant to remove an *invisible* pack from her pocket. She will then extract an invisible cigarette from it, giving a very neat and consistent appearance to the effect. The script will be almost identical to the non-smokers variation: simply add the word "invisible" when necessary. Just as before, interpretations may vary despite using the exact same words. In the "All in her hands" paragraph for example, you asked the participant to bring out her pack: the subsequent line "Oh, I'm not going to touch anything" will now be perceived as a gag (since you are not asking for a real pack this time) and cause the same chuckles obtained in the original version.

This version is fine for a suggestion-based effect; however, resorting to the chemical out would be far more difficult and require full audience control. I never tried it in these conditions. Instead of the Old Cowboy's last trick, I suppose you could drop the foil when pretending to remove the invisible pack from her hand, with the humorous excuses that she won't need it anymore and that she will be able to stub out the invisible cigarette on her palm more comfortably.

This approach would raise three problems: first of all, you would need to position the participant at a greater distance from the audience (in parlor/stage settings), to hide the foil from their view. The "curled fingers" subtlety would still be necessary unless performing on a heightened stage (or if you are sure that the rest of the spectators will obediently keep their seats).

Secondly, the spectator would need to keep her eyes closed until the climax, since the foil would not be covered by the pack. You could use the familiar concentrate/close eyes equivalence: if she doesn't close her eyes during the induction ("Sue, when people *concentrate* or focus they tend to visualize much better"), you will be able to provide another visual cue in the "Slow Center" phase ("Bring that cigarette closer to your right hand and *concentrate* on the center of your palm"). In the "Trust and responsiveness" paragraph, you wouldn't of course ask her to open her eyes ("Please try to stay awake"), but instead congratulate her as soon as you notice that she has closed them with something like:

> *VERY GOOD now, you really need to KEEP THAT CONCENTRATION UNTIL THE END as you continue feeling this way, okay?*

This very unspecific request would bring an additional "Yes" and hopefully have the spectator keep her eyes closed (i.e. "keep that concentration") until the end of the experiment. "VERY GOOD" should be said after she closed her eyes, and suggests that she did the right thing.

Lastly, the participant may feel something when you secretly drop the foil in her hand, despite the minimal weight and the short distance: the impact this time would not be covered by the larger and simultaneous brushing of the real pack and by the other distracting factors described earlier. You could perhaps put your right hand on her shoulder during the drop, for added misdirection.

This said, I never felt the need to try the chemical out/kicker in these conditions, because the presence of the pack in the original version is never questioned.

VI.3 Nostalgic variation

This is another no-pack variation, but with the foil left in full view. I performed it for several years with good results, before coming up with the "covert" approach. It is easier than the versions described earlier, and has no angle limitations, no minimum distance requirements and no lighting preconditions; however, its impact on the audience is not as strong. I report it here in case you wanted to become acquainted with the gimmicks and the suggestive patter in a simpler way. Of course, we will take this opportunity to explain a number of additional subtleties along the way.

1. Terminology

Proceed as in the original version until the "All in her hands" paragraph. When the spectator brings out her pack of cigarettes, add:

> *Could you please tear up a VERY LITTLE piece of the paper inside and keep it in the center of your right palm, because you're going to ignite it soon...*

She should tear the foil herself, because you obviously don't want to touch anything. Besides, if you have already set up the gimmick on your fingers, handling the foil may start the chemical process (and tearing it without using your thumb and index finger would definitely look unnatural!).

A mistake I made for years: asking to tear up a piece of "aluminum" or "foil". Call it just "paper": why adding detrimental details? Let them remember (and tell their friends) that she pretended to light a piece of *paper* in her hand. Incidentally, the foil inside the cigarettes packs is made of 50% aluminum foil and 50% regular paper: you can call it "paper" with the same rights as you would call it "aluminum foil". We stress out "very little" ("Could you please tear up a *very little* piece of the paper inside") because the participant generally rips a piece larger than necessary: even if we don't have visibility or weight concerns here, this would cause the chemical to spread on a larger surface, reducing its effectiveness. Besides, a "very little piece" suggests something innocent and insignificant. Having her tear it again would cost some extra time, but if necessary, ask her:

> *Please make it smaller, because you don't want to hurt yourself too seriously!*

To make sure she complies, we provide once again valid reasons for our commands: why does she have to tear a piece of paper and keep it in her hand? Because she is going to ignite it soon with her invisible cigarette. Why does she have to make it smaller? Because she[18] doesn't want to hurt herself too seriously.

2. Fire and forget

Once she has torn a piece of the right size and pretended to extract an invisible cigarette from the pack, continue with:

> ... and it is best if you PUT THE CIGARETTES AWAY AND FORGET ABOUT THEM, because you won't need them anymore, okay?

You certainly noticed the subtext and the embedded suggestions: the first meaning is to replace the pack in her pocket, because she won't need it during the experiment. The second is to put away the cigarettes and completely forget about them (i.e. forget about smoking), because she won't need them anymore after our hypnotic cure. The command links a conscious action ("Put the cigarettes away") to an unconscious action ("Forget about them"): this conscious/unconscious split is commonly applied in medical hypnosis[19], to make a suggestion to the subconscious mind (forget about the cigarettes) by associating it with an easy physical command (put the cigarettes away). This type of link is employed with alerting frequency throughout the book - please reread the scripts and find the right way of dealing with it.

---

[18] You got it, *she* doesn't want to hurt herself.

[19] For example, in phobias treatments: most of the time the subject consciously acknowledges the irrational nature of her phobia, but on a subconscious level the phobic response is triggered regardless.

Follow-up with the "Better now" line ("You feel better now, don't you?") as soon as she replaces the pack in her pocket. She feels better because you will not take her pack and make anything strange with it, and (on a deeper level) because forgetting about the cigarettes certainly has a positive impact on her health conditions...

3. Moves

Proceed with all the suggestions as per the previous versions, until the "Old Cowboy" paragraph; in case she wishes to "continue even further", or "to make it even easier", you will take the foil while saying:

> *Let's make it longer because you're going to ignite it like a fuse with your invisible cigarette...*

This excuse is very reasonable: since it is difficult to ignite a piece of paper laying flat on the palm, it is best to roll it like a fuse, L-shaped and bent upward (Figure 16), creating a meaning for the otherwise unjustified rubbing and folding actions.

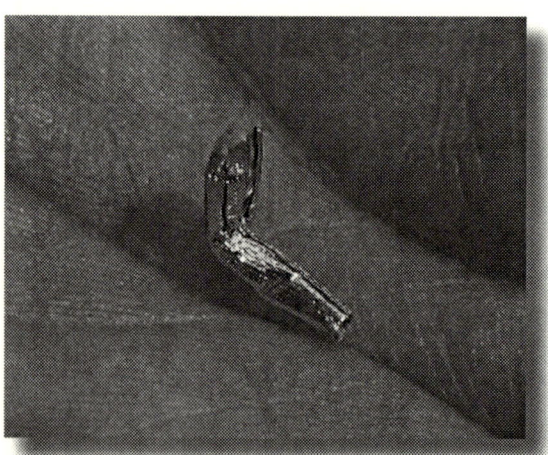

Figure 16

There are two methods for the move: the easiest way consists in having some Hypnoheat dust on your fingers and openly rub/fold her piece of paper into a fuse. In fact, the spectator is unlikely to notice any rubbing action since the paper will be almost entirely covered by your fingers: she will only see you make it into the proper shape, as anticipated. Be sure to casually show your hands empty both before and after the move.

Alternatively, you could ask the participant to fold her paper herself, while you secretly prepare the gimmicked foil in your left pocket (just like in the original version). You will then switch them with the excuse of quickly reshaping her rolled piece. The move is borrowed from Tony Corinda's *Thirteen Steps to Mentalism*: reach for the spectator's paper with your right thumb and index finger, while your left hand comes out of your pocket, secretly holding the prepared foil in pinch grip. Remember to move your right hand earlier and faster than the left for misdirection purposes. Pretend to reshape her piece of paper, with the thumbs and first fingers of both hands (Figure 17, shown with oversized, colored pieces of paper for greater clarity).

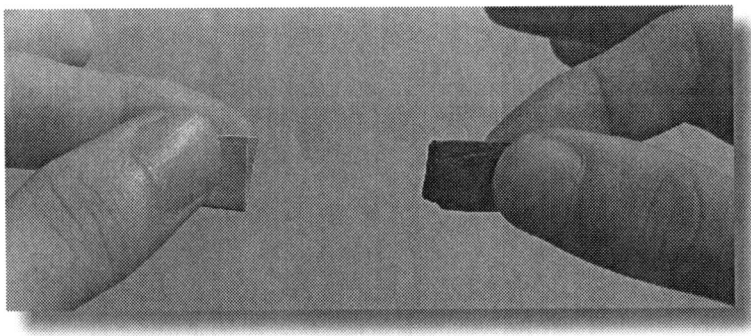

Figure 17

The left fingers then push and conceal her paper between the right fingers (Figure 18), while the prepared foil is now the one held in full view (Figure 19). For added misdirection during the switch, you could ask her to replace her right hand palm up, in its initial position. The switch is virtually invisible: the paper is in your hands for less than 5 seconds and is hidden from her sight for just an instant: few laymen will ever remember that you touched it (the audience behind you may not even see it). As opposed to a billet peek, where the importance of the written information may have well-versed spectators remember that you touched the billet even only for a second, the brief reshaping of that torn paper is irrelevant and will pass completely unnoticed.

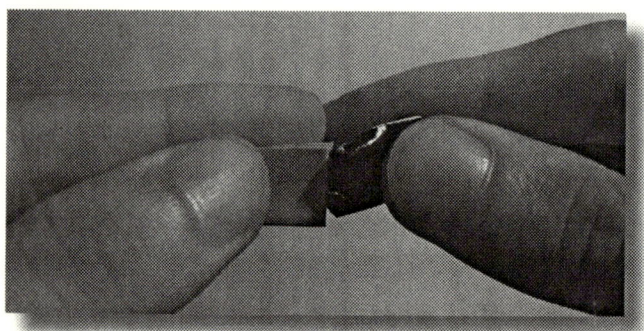

Figure 18

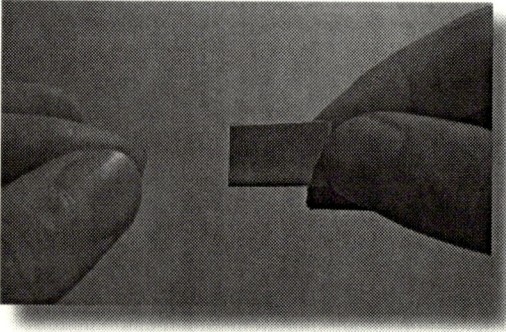

Figure 19

The participant will then apparently ignite the fuse with her imaginary lighted cigarette ("whenever she wants"), and feel the burning sensation as in the original version. In the "Burning now" paragraph, you will obviously substitute "Carefully stub it out on your hand" with "Carefully ignite that fuse in your hand".

4. Applied misremembering

Conclude the effect as per the original script. You will have to adapt the misremembering patter accordingly, by replacing "When you finally decided to stub it out on your hand" with "When you finally decided to ignite that fuse in your hand", and by adding:

> *You tore up a very little piece of paper and simply kept it in the center of your right palm - you did that didn't you?*

You probably noticed that some of the misremembering tools explained earlier were put into good use. For example:

- We didn't mention that her piece of paper was folded into a different shape (let alone that you had a part in it): it was just a utility move and seems irrelevant - none will ever argue with you for this omission.
- We used the same words ("You tore up a *very little piece of paper*", "And simply kept it in the *center of your right palm*", *"Ignite that fuse* in your hand").
- We added the extra details discussed earlier (the center of her right palm).

5. Conclusion

The "fuse" metaphor may not be obvious, but smoking is often compared to a time bomb ready to explode in your body: you don't feel its consequences for a long time, and then suddenly pay a high price.

Although less surprising than the previous versions, this variation still gets excellent reactions. Of course you won't be able to perform the original effect for a crowd who has already witnessed this beginner's version, since the spectators would naturally suspect the use of a hidden foil.

VII.   AFTER MINT: AUDIENCE ASSUMPTIONS

While experienced mentalists sometimes think about mercury, laymen hardly suspect any trickery at all, unless you are a magician preceded by your reputation – this is why you tried to suggest throughout the routine that you weren't touching anything at all. They believe that the effect was achieved through hypnosis and suggestions, which is partially (or completely) true, and may talk about trance, altered states etc... You can bring the conversation into your favorite areas from here, but the most popular discussion topics consist in the quit-smoking methods. You are presenting yourself as an expert in mind matters with a specific expertise in smoke addiction, and you must be aware of the common tools employed to fight it. For example, a spectator asking about aversion therapies deserves much more than an embarrassed smile, especially because this type of technique was used in the experiment.

Some of the methods described below are valid, others seem a bit "off" – but who am I to judge... I don't even smoke.

VII.1   Aversion techniques

The **Aversion Therapy**, which has roots in the Behaviorism school of Psychology, consists in reminding the subject of all the things she dislikes about her addiction: it works by linking the unwanted habit to unpleasant sensations, so that she becomes conditioned to repulse it. In our case, the main negative elements would be cough, sour throat, bad breath, yellow and decayed teeth, stinking clothes, skin burns and ashes. There are many variations: for example, the **Rubber Band Method** is used to stop the flow of unwanted thoughts. The subject needs to wear a rubber band around her wrist, and ping it violently (Figure 20) every time she is seized by the desire of smoking. Consequently, the unwanted thought becomes unconsciously associated with the pain and is gradually eradicated. Another advantage is that you will always be ready to perform some Dan Harlan's tricks.

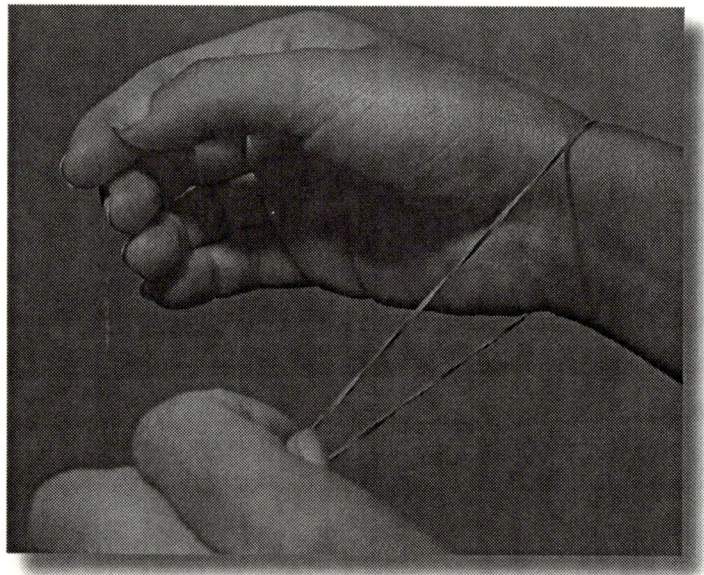

Figure 20

Along the same lines, but with a harsher approach than our metaphoric performance, comes the **Brutal Hypnosis**: the subject is asked to imagine that she is actually experiencing the worst possible consequences of smoking, for example dying of lung cancer! Many shy away from this idea, and try to avoid such cruel suggestions. But maybe the end justifies the means... This is a popular belief these days, one just needs to look at the pictures of the new European Union's anti-smoking crusade (Figure 21). This 90-million-dollar media campaign calls on EU member states to put shocking pictures on cigarettes packs: it features 42 photographs, including decayed teeth, rotten lungs, a newborn with an oxygen mask, and some "subtle" allusions – such as a wrinkled apple symbolizing premature skin ageing, and a bent cigarette representing the loss of male fertility. Talk about vivid imagery[20]!

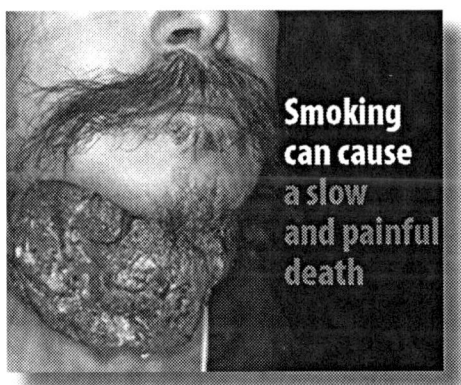

Figure 21

---

[20] During a press conference held in Brussels on 22 October 2004, David Byrne, EU Commissioner for Health and Consumer Protection, justified the campaign with the following: "People need to be shocked out of their complacency about tobacco. I make no apology for some of the pictures we are using". Smokers' lobby groups immediately retorted that written and graphic health warnings should also be displayed on alcoholic beverages... Fortunately there are no reports of secondhand alcohol. People should be free to kill themselves in the way they fancy the most, as long as they don't hurt anybody else in the process.

The **Rapid Smoking** method was described by Judy Perlmutter in *Stop Smoking in Five Days*: at scheduled times during the day, the subject over-smokes in a way that makes the cigarettes taste unpleasant, and repeats discouraging phrases such as "smoking burns my lungs", "smoking irritates my throat" etc.

Just like everything else in life, the aversion techniques benefit from gimmicks, such as special breath sprays or mouthwashes to be used before smoking: they alter the taste of cigarettes (making them nauseating), without affecting regular food. Once again, the objective is to associate an unpleasant sensation with the smoking impulse. It reminds me of Anthony Burgess/Stanley Kubrick's *A Clockwork Orange*, where doctors force the main character to watch violent movies during painful medical treatments, in order to cure his aggressive behavior.

VII.2   Other psychological methods

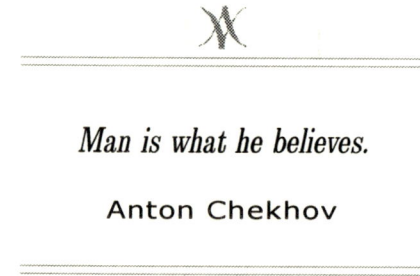

Man is what he believes.

Anton Chekhov

Self-Hypnosis (a.k.a **Autosuggestion** or **Autogenous Training**), is a process by which an individual trains her subconscious mind to believe in ad-hoc suggestions, or systematically schematizes her own mental associations for given purposes. For example, the subject can reach an altered state by focusing on herself while listening to trance-inducing audio tapes: the scripts address specific issues such as weight reduction, stress management, panic attacks etc. and generally rely on guided imagery and relaxing background music. Thanks to a special arrangement with Jef Gazley, a certified hypnotherapist from Arizona, you can download a free sample of his "Smoking Cessation" CD (in MP3 format) from the following Internet address:

http://www.asktheinternettherapist.com/samples/
Smoking-Cessation-Hypnosis-CD-Sample.mp3

Manifesting is a mental practice consisting in creating your own reality through the use of visualization techniques[21], positive thinking, self-analysis and consistent self-deception. The subject willing to quit smoking, for example, would need to visualize (hence, recreate) herself as a non-smoker. To a certain extent, we are what we define.

The **Cold Turkey** variation uses sheer willpower, without any drugs or gimmicks. After setting the rules on when and where to smoke, the subject begins to eliminate the cigarettes she enjoys the least, until she only keeps the ones she needs the most. Many people try to stop by telling themselves that they don't want to smoke anymore, but they generally don't believe themselves.

---

[21] This is the word we repeatedly used during the induction, when asking the participant to "visualize" the scene ("You can begin to relax deeper and try to *visualize* it in your hand", "When people concentrate or focus they tend to *visualize* much better", "When you *visualize* that white, long cigarette").

Through the **Gradual Cut-down**, the subject can either reduce the number of cigarettes smoked during the day, or lower the amount of nicotine absorbed into her body. Of course the second method requires a cigarettes filter: it usually looks like a small plastic holder (similar to a pull), which fits on the mouth end of the cigarette and traps some of its harmful chemicals.

Finally, the **Primal Scream** therapy applied to smoke addiction consists in screaming whenever you feel the desire of smoking, in order to repulse it. This theory derivates from Arthur Janov's "Primal Therapy" and Daniel Casriel's "New Identity Process" (a few years later, Nolan Saltzman suggested a similar approach called "Bio-Scream Psychotherapy"). While your neighbors may not always agree, yelling to evacuate stress can make sense - somewhat like taking a deep breath before walking on stage.

VII.3   NRT or NLP?

The **Nicotine Replacement Therapy (NRT)** works by gradually substituting the nicotine with milder drugs, in order to break the smoking habit without suffering from the well-known withdrawal symptoms (depression, anxiety, irritability, inability to concentrate). For example, Zyban is an anti-depressant drug alleviating some of those symptoms. Due to a number of side effects (such as insomnia and headache), it was supposed to be available upon medical prescription only, but is now easily purchasable over the Internet. Another NRT drug, called Rimonabant, targets simultaneously hunger and cigarettes craving. Since the hunger signals and the nicotine craving are hard to tell apart, people often smoke instead of eating and tend to take weight when trying to quit smoking, creating a vicious circle: when the subject smokes instead of eating (confusing hunger with nicotine craving), the blood sugar levels fall, due to the missed meal. This increases the carbohydrates craving, causing stronger hunger signals that may again be misinterpreted as cigarettes craving etc... Keeping the blood sugar levels even during the day can help reducing the number and severity of cravings, and consequently cut down the number of cigarettes smoked. Eating regularly would be the most obvious solution, but it is not always practical - hence the idea of a **Glucose Treatment** (usually in the form of dextrose tablets) aimed at raising the blood sugar levels. Glucose may also help fighting the nicotine withdrawal symptoms by boosting the production of serotonin, one of the brain's "feel-good" chemical.

Homeopathy originated in the nineteenth century with Samuel Hahnemann (!): this medical practice uses small quantities of remedies that in larger doses would produce effects similar to those of the disease being treated. It is based on the principles that "Like cures like" (fight the disease with its own weapons) and "Less cures more" (dilute the medicine to increase its effectiveness). Homeopathic remedies against smoke addiction include patches, chewing-gums, syrups etc.

VII.4   The New Age approach

In addition to the standard medical practices and psychological treatments described above, a number of New Age psychotherapies have flourished in recent years. Many of them are based on traditional techniques and adapted to modern requirements.

Aromatherapy was founded about one century ago by French doctor René Gattefossé, and uses the smell of essential oils extracted from roots, seeds, plants and flowers. There are a number of "stop-smoking" perfumes sold in New Age shops – you never know where the solution might come from...

Auriculotherapy, a variation of traditional acupuncture[22], is based on the belief that the ear is a map of the bodily organs: for example, a smoke-damaged lung should be treated by sticking a needle into the part of the ear corresponding to the lung. The **Laser Therapy** is a high-tech version of this technique, using a painless laser beam instead of the frightening needle.

Crystals, gems and other precious stones have always been admired for their beauty and alleged mystical powers: the **Gem Therapy** pretends that crystals can actually harness and unleash positive energy against diseases, injuries, addictions etc. This theory is based on a property of certain crystals, which produce an electrical charge when compressed (the so-called "piezoelectric effect").

The **Urine Therapy** is self-explanatory: this age-old practice requires you to drink your urine. Homer Smith (*Man and his Gods*) wrote that "Man is a machine / for turning wine / into urine". Many advocates claim that urine is a panacea, effective against all ailments: toothache, snakebites, lungs infections, sour throat etc. Imagine the magic ad: no added bulkiness, perfect for walk-around situations, free refills.

---

[22] Acupuncture is a Chinese medical technique consisting in sticking needles into specific points of the body, to unblock obstructed pathways of "Chi" (the vital energy believed to flow in all human beings).

Ear Candling (or "**Coning**") consists in cleaning the ears and the mind from negative energy and residual poisons, by inserting a lit hollowed cone into the ear: this cone will allegedly suck out the harmful chemicals retained by your body. Do not try this at home!!

The list is endless... Joel Wallach (*Dead Doctors Don't Lie*) pretended that most health problems were due to mineral deficiencies, and suggested consuming colloidal minerals to solve them. Some marketed items blend multiple techniques: for example, the "ZeroSmoke" method combines the auriculotherapy (as we have seen, a variant in itself of acupuncture) and the gem therapy (in this case, the nobility of gold and the property of magnets). Basically, it's a pair of magnetic golden earrings. According to the ad, those special earrings stimulate specific points of the ears, supposedly causing neural transmitters to produce the same endomorphins that the body generates when inhaling nicotine.

Enough research will tend
to support your theory.

Arthur Bloch

VIII.   THE HYPNOTIST'S STANDPOINT

The scripts described so far were fast-paced and aimed at real close-up situations, where any effect lasting more than a few minutes would probably result in grumbles of impatience from the audience. Other settings, such as pseudo-hypnotherapy sessions or intimate performances for that special person, may instead grant us a longer time framework (and the use of cribs). This chapter provides further examples and subtleties for those situations.

Before analyzing a longer "warm-up" and a full quit-smoking session transcript, let us highlight an often-overlooked tool to enhance rapport with the participant: adjusting the hypnotic induction to the subject's typology and temperament. Similarly to how we adapted the presentation to the type of audience (cure for smokers vs. prevention for non-smokers), the induction should reflect the personality type of the subject, and the way she is feeling in that specific moment. For example, a subject in a "hyper" state should initially be addressed in a quick and brisk tone, mirroring her emotional condition and naturally involving her subconscious mind (hence, creating greater rapport). Dr. Maurice Kouguell, director of the Brookside Center for Counseling and Hypnotherapy in New York, kindly provided the following classification.

VIII.1   Scripts adjustment (by Maurice Kouguell)

1. For the Introvert

"The Introvert is characterized by a hesitant, reflective, retiring nature that keeps itself to itself, shrinks from objects; is always slightly on the defensive and prefers to hide behind mistrustful scrutiny". (Carl Jung)

For such a type the induction will acknowledge their needs. Let them be introverted without judgment. It is suggested that a familiar trance experience be used. Always respecting and allowing their need for privacy and time. Introverts may experience trance very frequently, for they spend a good deal of time focusing on their inner world. It is also wise to remind them that their need for privacy will be respected.

2. For the Extrovert

"Extroverts are normally characterized by an ongoing candid and accommodating nature that adapts easily to a given situation, quickly form attachments and setting aside any possible misgiving will often venture forth with careless confidence into unknown situations". (Carl Jung)

For these clients, it is suggested that the induction be congruent with their life experience, possibly focusing on the person or situation with which the extrovert is preoccupied. The induction might take the way of beginning with a familiar process and gradually adding on the extrovert's life experience to it. Encourage them to feel that they will find answers within themselves.

3. Sensing type

"Those with a sensing preference choose to use their physical senses as the major source for information gathering. Sensing types tend to be quite realistic, practical and present orientated, since what they know is based on what they have seen, smelled, heard or touched". (Marie-Louise von Franz)

For the sensing type, it is suggested that attention should be paid to details of the trance rather than to the overall experience and eliciting sensations based on previous experiences.

4. Intuitive type

"This type focuses on global aspects and can relate best by proceeding from the general to the specific. Metaphors, symbols and abstractions are welcomed by that client. Introverted intuitive types may appear to be odd, eccentric and preoccupied with daydreaming and are not deeply concerned about being understood by others. Extroverted intuitive types are guided by hunches, rather than facts, may vacillate from one idea to the next and may be seen as creative and flexible". (Baldwin R. Hergenhahn)

The approach of the induction should include looking at future possibilities while using techniques of story telling, metaphors and symbols. Since daydreaming and imagination are their characteristics, those can be used.

5. Thinking type

In this type, the individuals have a need for control and they need to be made to feel that control is acceptable; that in time, they will accept that they have the control over giving up control. It may be wise to ask them what imagery they might like to visualize. The process of induction should take into account the above and allow them to develop their own style of going into trance.

6. Feeling type

Those individuals take decisions based on values or emotions without necessarily resorting to logic. The induction needs to focus on specific feelings. It is suggested that past experiences will facilitate entering a hypnotic state, and allow the client to feel accepted.

Those are examples of how understanding temperaments and types can be a helpful adjunct to structuring an induction. As illustrated above, each preference has its own characteristics; acknowledging and recognizing them will certainly enhance rapport.

VIII.2   General induction (by Mark Tyrrell)

With the basic typologies in mind, here is an example of longer induction with a warming theme. The script, which applies many of the tools described previously, is taken from the audiotape *How to Use Hypnosis to Help Yourself and Others*, by British hypnotist Mark Tyrrell of Uncommon Knowledge Ltd.

KEY:

Underline = embedded suggestion, *Italic* = dependent suggestion[23], **Bold lower cases** = presupposition, CAPITALS = nominalization[24], **BOLD CAPITALS** = illusory choice, Lighter fonts = conscious/unconscious split.

"Okay, now, you can go into hypnosis **WITH EYES OPEN OR EYES CLOSED**, but it may well be more COMFORTABLE just to take a moment to close your eyes right now. An interesting thing is that **when you begin to relax deeply**, the flow of blood in the body is altered – when a person becomes tense, blood tends to leave the stomach and go into the major muscle areas and people can develop digestive problems but when you RELAX, quite often parts of the body feel warmer. Sometimes the hands feel warmer and blood flows into the hands and the stomach often begins to function in a very NICE, EVEN way **as you begin to relax.**

---

[23] Some of you may be left wondering about the difference between an "embedded" suggestion and a "dependent" suggestion. For those who care, embedded suggestions (commands) are hidden orders to the subconscious mind: for example, "When you relax, quite often parts of your body feel warmer". Dependent suggestions are instead relying on other suggestions: for example, "As you become aware of those warmer sensations (first suggestion), you may begin to *feel much more comfortable without that cigarette* (dependent suggestion)".

[24] The capitalized words in this induction indicate the nominalizations (some of them are not capitalized to avoid confusion with other language patterns), while capitals were previously used to mark important keywords, through louder tonality and direct eye contact with the participant.

Now what you can do is just to take a few seconds now to imagine the sort of place where you could be at this time on listening to this you are BEAUTIFULLY, NICELY, PEACEFULLY RELAXED, the sort of place where you can really give the space in your mind, where you can really learn and discover the new ways of doing in your life. Now in a few moments I am going to count from one to ten and you can just allow the process of listening to those counts to take you more and more into that place noticing the things you could see and hear or taste or smell and just be aware of once being in that SPECIAL RELAXING place. I do not need to know where that is and you may not even know where that is **until you find yourself there**. It could be **A FOREST OR A BEACH, FIELD, SOME PEOPLE TALK ABOUT WATERFALLS, MOUNTAIN TOPS, VALLEYS. IT COULD BE ANYWHERE** you find particularly PEACEFUL and RESTFUL and **as you become more aware of this place** *you can begin to notice changes in you.*

So from 1 **just becoming more and more aware of how you breath deeper** as you RELAX more and how you can notice temperature variations *in different parts of the body* **as you go down** to 2. And now you can imagine yourself walking down to that place through **A FIELD OR A PATH OR A ROAD OR STEP OR JUST DRIFTING THROUGH** to that place, that's it, 3... RELAXING deeper and... you know the more you experience CALMNESS in your everyday life, the more clearly the mind can work in certain ways. And 4... that's it, drifting **DOWN AND ACROSS OR UP** to 5. Just becoming aware of the sort of shapes you would be able to be aware of. **BIRDS**

SINGING OR WATER OR WHATEVER you would experience in this special place... you know, the skin on your face may alter color just a little bit the color as blood comes to the surface. You know when people are very tense they tend to go pale some times but **when they relax**, blood can come to the surface and move around the body more freely. You RELAX more and more to 6... and 7 and you know when you go into hypnosis, *the mind wanders inward* the same way that it does when you dream sleep. *The mind wanders inward* and you can travel the expanse of your CREATIVE inner reality. That's it... 7,8 that's it just drifting through... 9 and you can begin to prepare to drift down to the next number and **when you do so** *you can notice how relaxation can extend more and more* throughout your body drifting around the mind. Part of you can do what it likes and part of you can be aware what of real REST[25]... and 10 just drifting through... you don't have to see or be aware at all levels of the SPECIAL place but a part of you, even the part you are not aware of, can be aware of this SPECIAL place... and the PEACEFUL and SERENE effects that this place can have for you and on you. Almost as if just to sit in this place can allow muscles to REST in the neck and back the legs and the arms and even the bones to REST, almost as if they are covered by an invisible quilt of TRANQUILITY... that's good... and all the organs to REST and the joints..."

---

[25] Please note the extensive use of confusional language at the end of the induction.

VIII.3 Session trance-script (by Nicola Dexter)

As we continue past the induction phase, here is the transcript of a quit-smoking hypnotherapeutic session by Nicola Dexter, a professional hypnotherapist from London, England who developed a unique way of supporting people in stopping smoking and generously shared her work. Her script is broken down into small sections: she would first ask specific questions about the subject's particular circumstances, then mix and match the sections depending on the need of the client. You will recognize many techniques described in the previous chapters, such as the involvement of the subject's five senses, the use of embedded commands and illusory choices etc. You have the tools to uncover the deceptions hidden below the words and employ them in the most constructive way.

1. General

Because you are choosing to relax today and you are choosing to **take control of your life**, you know that now is the time for you to **stop smoking**. You know the difference between when you are just paying lip service to something and when you say something that you really mean, and you know that you really meant it when you said to yourself that **you want to stop smoking** and that **you are going to stop smoking today**! Or maybe you even said to yourself that **you really need to stop smoking** and that **you are going to stop smoking today**. You wouldn't even be here today if you really didn't mean to **stop smoking today**. During this special state of relaxation, your subconscious mind becomes more open, so that it can accept ideas, suggestions and concepts that are for your benefit... these suggestions will take root in your subconscious, and nothing that is accepted into the subconscious is easily removed, so that you can be certain that they will be there, helping you, guiding you, over the coming days, weeks and months, helping you to **stop the smoking habit** easily and completely... helping you to **stop the smoking habit** for good... so that something that you thought might not have been easy to do becomes much easier than you could have imagined before today... and you will take a great deal of pride and pleasure in the ease with which you **stop the old habit** for good...

2. Dealing with others around

Your friends and relatives will be surprised too, and if they smoke themselves, they are going to be none too pleased at being reminded that they still have a habit which **you have freed yourself** from so easily. And if they don't smoke but have a different addiction, they may be slightly envious that **you have dealt with an addiction** so powerfully when they are still stuck with theirs. And if others around you happen to offer you or urge you to have a cigar or cigarette, you'll simply find yourself smiling and saying "No, thanks, I don't smoke" or "No, thanks, I don't smoke anymore"... You'll be amazed at just how easily you can say those words "No, thanks, I don't smoke". You'll just find yourself saying it without any effort at all and you're going to find yourself with a little surge of pride and pleasure at being able to truthfully say "I don't smoke". So, something that some people say is difficult and needs willpower, you will find is easy when the subconscious and conscious parts of your mind, and the automatic systems of your body, begin to work together for your benefit... helping to make it easy for you to **stop the smoking habit** easily and completely...

3. Dealing with reasons client started to smoke

Now I would like you to imagine a blackboard, and on that blackboard you are going to write down the reasons why you started to smoke or how you have justified continuing to smoke. I will make some suggestions and if that reason applies to you then write it down:

- Maybe you started to smoke due to peer pressure but you know now that you would much rather **make your own choices, be your own person** and that your friends only wanted you to join in with them because they were too weak to say "No", and if a friend would have stopped being your friend because **you refused a cigarette** then they were not really the type of friend you would want...
- Maybe you started because it seemed to be a grown up thing to do but you are grown up now so you no longer need anything to make you seem more grown up...
- Maybe you thought it was a cool image – you saw film stars smoking but you know now that they didn't show the negative side of the film stars smoking, coughing, others recoiling due to the smell of their bad breath, smelly clothes and fingers...
- Maybe you thought that it helps in slimming but you know now that the body can only survive for so long on cigarettes and insufficient nutrition before it starts to get diseases or a lack of energy...
- Maybe you thought that it has a calming influence but that was only because of the deep breaths you took as you inhaled, so now you can take deep breaths to calm yourself but without the cigarette...

- Maybe you started to smoke to be rebellious but who are you rebelling against now...
- Maybe you have been smoking to give yourself comfort but do you really need a dummy substitute...

So, when you have written down all those reasons why you smoke and any more that are specific to you, I would like you to look at the blackboard and then pick up a blackboard rubber and start to rub out those reasons, and whilst you are doing so you are sending a powerful message to your subconscious mind that **you no longer wish to smoke** for any of those reasons, **you no longer want to smoke** for any of those reasons and **you no longer need to smoke** for any of those reasons because **you have chosen to be a non-smoker** so none of those reasons are valid anymore...

4. Getting client to rewrite history

Now I would like you to imagine that all your past stretches out behind you in a long line – your time line – and I would like you to imagine floating up above your time line and turning round so that you are facing the same direction as your past. I would like you to float all the way back over your time line to the point where you first chose to have a cigarette. Picture yourself there, where were you, who were you with, how had you got the cigarettes and then I would like you to float down and speak to that younger you with the knowledge you have now and communicate with that younger you until he/she no longer wants that first cigarette... Then float up above and see that younger you either throwing the cigarette on the ground or saying "No thanks" or simply ignoring the cigarettes available... And then when you have seen that happen start floating all the way back to now, realising that if starting smoking had only taken a single decision which you have now changed, then stopping smoking needs only a single decision to take your life on a different path, one where you have chosen a healthier path, one where you have chosen to **be a non-smoker** and by the time you get back to now you realise that this really is all it takes - one single decision...

5. Dealing with client's disempowering beliefs

Many people say that they just don't have the willpower to **give up smoking** but what really is willpower - if we cut you open could we find it... The reason most people think that they don't have the willpower is that they are trying to use their conscious mind to overpower their unconscious mind. The unconscious mind is incredibly powerful and takes control of all your automatic behaviours, however it speaks a different language - it speaks the language of visual images so if you want it to behave differently you need to picture the new behaviour and not simply speak it. So picture an image of you smoking a cigarette, smoke all around you, coughing, maybe imagine the image as dim, unappealing, maybe in black and white! And then picture a large, bright, colourful image of you without a cigarette, with a smile on your face, clean air around! And then put a big cross over the first image and a big tick over the second image. You have now spoken the language of the unconscious mind. Now with regards to willpower, **you can achieve what you can picture**!

And if you want to get yourself to take the actions you really want to take, you need to become a person who is their word and not their feelings. We can all make excuses about why we have not kept our word, but when we practice honouring our word and only promising what is authentic and then keeping our promises, our lives will be much happier. Just as your life will be much happier when you have kept your word to yourself that **you no longer smoke**. You now imagine yourself being a man/woman of your word – people can really count on you and you can count on yourself and **you become a non-smoker**.

6. Dealing with client's concerns about eating more

If you are concerned about eating more as **you quit smoking**, then firstly you need to decide what you would really like to eat on a regular basis. Choose what you would like to buy from the supermarket or shops before you get there. Picture the items you would like to buy, maybe write a list and see yourself going round the supermarket or shops – stopping at the fruit and vegetables and buying more of them, walking past the unhealthy foods such as cakes, sweets, chocolate[26], you simply walk past those aisles because **you are now committed to eating and drinking more healthily.**

Picture three tables – the right one with all the items you can eat lots of, the middle table with all the things that you would just like to eat occasionally and the left hand table with all the things that you really do not want to buy or eat. Sweets, chocolate, processed foods, things with sugar in them, cakes and biscuits, things with artificial sweeteners in, put them all on the left hand table. On the middle table put things like meat and fish and maybe bread and dairy products, and on the right

---

[26] Depending on the context, you may want to mention alcohol and unspecific fat/processed junkfood in the "forbidden food" category, instead of chocolate, cakes and sweets (I would be afraid of losing rapport)...

hand table put fruit, vegetables, nuts, seeds, sprouting seeds, water, fruit and herbal teas etc... Then push all the things on the left hand table into a big black bin - they are all the things with empty calories, no nutritional value or poisonous to your body.

Your body is a delicate organic vehicle which historically has been used to fresh, raw food, and which in the space of less than a century has been given food that it clearly cannot digest. Is it any wonder that our bodies keep getting sick with the rubbish that they are fed? If you put the wrong fuel in a car you would expect it not to work properly, the same applies to your body...

7. Dealing with smoking due to emotions

Some people say that they smoke because they are bored - I ask how can an adult be bored with the infinite number of choices we can make as to what to do? All being bored means is that you haven't bothered to choose what you could do instead of nothing - you could read, you could write, you could talk to someone, you could find a hobby, you could listen to music, the radio, watch the television, invite friends round, go to the cinema, cook a healthy meal, learn a skill/language/art, you could visit a museum, take a walk, go to the theatre, restaurant, library, you could find information on a computer or email friends, play a game... There are many, many possibilities of what you could do - as an adult you need to grow up to be self-generative and if you didn't get taught to do that then learn...

You may smoke using the excuse that you are stressed. How do you get yourself stressed as the same situation on the outside may cause some people to get stressed and not others - choose the situation you are in, rather than resisting it, focus on how you would like things to be rather than on how you do not want them to be - remembering that your unconscious mind moves towards the images you have in your mind. Picture yourself being calm, relaxed, confidently dealing with everything that needs to be dealt with and picture yourself drinking plenty of water, and see yourself without cigarettes but breathing deeply whenever you want to feel calmer...

8. Dealing with withdrawal symptoms

When you draw on a cigarette the body responds to the nicotine taken into the body by producing chemicals which is what you crave. I would like to ask your unconscious mind to recall all the chemicals that are produced when you smoke a cigarette and I would like to ask it to keep on producing those chemicals after **you have stopped smoking.** And I would like your unconscious mind to continue to create those chemicals for the next four weeks but gradually cutting them down until they are no longer being produced in four weeks time. And I would like to let you know consciously that by doing that it will feel as though you were just gradually cutting down smoking, and your body will feel much more comfortable and you will be most surprised at how good you feel once **you have stopped smoking** and made use of the amazing power of your unconscious mind to support you in stopping smoking...

9. Associating client with negative consequences

I would like you to recall images of people you have either known or seen who have been hooked on smoking and have become ill – hear them coughing, their lungs trying desperately to cough out the tar and nicotine that is clogging them up. Hear the wheezing as they desperately try to take in oxygen into their lungs. Smell the disgusting smell of cigarettes on their clothes, in their hair and recoil as you smell their breath. See all the lines around their eyes, around their lips and maybe even look at their fingers to see them stained brown from nicotine. Imagine their arteries being clogged up so that the blood can hardly get through, knowing that they have high blood pressure as the heart desperately tries to pump the blood through smaller and smaller tubes. Recall people who have died of cancer and emphysema. Smoking really can lead to these effects – it is no good living in denial – you are poisoning your body every time you smoke and one day it may be too late if you don't **stop smoking now**.

10. Metaphorical representation of choosing to stop

I would like you to imagine that in your mind you have a switchboard – a bit like the old telephone switchboards with plugs in it and wires from the plugs. Your switchboard is connected with smoking, so that all the plugs that are plugged in represent all the different reasons why **you used to smoke**. Attached to the wires are labels marked with the different reasons – so there might be labels like "first cigarette of the day", "cigarette whilst on the telephone", "cigarette after a meal", "cigarette with a coffee/tea", "cigarette whilst watching TV", "cigarette whilst driving", "cigarette after sex", "cigarette with an alcoholic drink" etc. What I would like you to do is to one by one look at these plugs and then choose to pull them one at a time. As you pull each one out I would like you to either think or say out loud "I choose to **give up**

smoking the first cigarette of the day" or "I choose to **stop smoking** whilst watching TV" or "I choose to **stop smoking** for whatever the reason on the label". If there are any plugs left in with blank labels then yank those out saying "I choose to **stop smoking** for any unconscious reasons, I choose to **be a non-smoker** now". When you have completed pulling every last plug out then please simply nod your head... then destroy the switchboard irreparably whether smashing it with an axe or setting it on fire or however you would like to do so whilst saying to yourself "I choose to **be a non-smoker** now"... and again when you have destroyed the switchboard then nod your head...

11. Setting up rewards for stopping smoking

I would really like to acknowledge your unconscious mind for supporting you in **being a non-smoker**. However, your unconscious mind also likes rewards. So, I would like you to allow your unconscious mind to come up with three rewards it would like to receive to thank it for supporting you in **giving up smoking**. These rewards need to be things that are ecologically sound for your mind, body and wallet. It could be things like having a massage or a beauty treatment, a night out, a weekend away or a holiday with the money you have saved from stopping smoking. It could be just keeping promises you have made to yourself to treat yourself that you have never kept. It could be buying something special - an item of clothing, a DVD, something for the computer, a CD, something for your home or car, it could be a trip to the hairdresser, it could be to have your teeth cleaned and polished. It could just simply be to take a long, hot, relaxing bath once a week or it could be to play a sport, take up a hobby or see a non-smoking friend.

So, I would like to ask your unconscious mind to come up with these three rewards in the next minute and when you have your three rewards then simply nod your head...

12. Metaphorical representation of cleaning the body

Now that **you have given up smoking** you need to think about the effects smoking had had on your body. I would like you to imagine that you could go inside your lungs to clean them out. So, first of all, imagine being inside your right lung and you have a hose and a scrubbing brush and I would like you to scrub the right lung clean, so that all the tar and nicotine breaks down into tiny pieces and is washed away with the water from the hose. And all that dirty water then gets eliminated from the body through the usual methods... then I would like you to scrub clean the tubes leading from the lungs to the nose and wash away all the tar and nicotine that has accumulated there as well. Then go down into the left lung and completely clean the left lung and when you have completed all that then simply nod your head...

Although not for everybody, smoking sometimes damages the nervous system and the spine so imagine that you could check the central nervous system and the spine and all the damage that has been done by smoking, imagine repairing it in whatever way seems appropriate so that it is all functioning properly again...

Also it is well known that smoking clogs up the arteries and veins. So, I would like you to imagine that you could go all the way round the circulatory system, through all the arteries and veins with a tiny pipe cleaner and as you twizzle this pipe cleaner what it does is any particles that are stuck to the artery or vein walls are broken into tiny pieces and get transported round in the blood until they can be excreted by the body. So, imagine cleaning the heart and every artery and vein in this way...

13. Future orientation

And, now that **you have stopped smoking**, promised yourself rewards and cleaned and repaired your body, I would like you to imagine floating forwards over your future and taking a look. See yourself going through your day, looking happy, without cigarettes, see yourself in your home without cigarettes, see yourself travelling without cigarettes, see yourself at work without cigarettes, see yourself socialising without cigarettes. Feel how proud you are of having given up. Hear the praises of other people who are impressed **you have given up**. Smell flowers and food. Notice how much nicer your food and drink taste. And then float over your future to 1 month from now - seeing yourself without cigarettes, then go to 3 months from now, see yourself without cigarettes, go to above 6 months from now, seeing yourself without cigarettes. Go to above 1 year from now, seeing yourself without cigarettes. Go to above 3 years from now; see how happy you are, maybe you look fitter, healthier, see how happy you are and how proud you are. And when you are ready, float all the way back to above now, drop down into now, back into your body and whenever you are ready in the next ten seconds or so then it is OK to open your eyes becoming wide awake. Wide awake and proud to **be a non-smoker**...

L'ENVOI

Many do not dare performing pure suggestive effects for the fear of failure. In this case, you have a "surefire" out (which is a strong trick in the first place) and simply cannot fail: even if the outcomes of the suggestions are disappointing, and the audience mocks you cruelly, you will have a winner in the end. It's easy to perform with confidence (and confidence is essential in these hypnotic routines) when the climax can't go wrong. Believe firmly in the effectiveness of your suggestions, and the experiment will be fully successful.

I performed the effect for well over a decade, and can vouch for the very strong impact it always carries: it can be used under any conditions, from a David Blaine's impromptu street magic show to an Ormond McGill's stage hypnosis act. It combines hidden gimmicks, secret preparations, physical manipulations, low-fat double-talks, hypnotic patterns, psychological subtleties and a meaningful, audience-centered presentation to create a killer routine – one you will be asked to perform over and over again. Repeated performances are a particularly interesting issue: people who have already witnessed the experiment (or who have simply heard about it) will be more responsive to the suggestions, due to the previous successes you obtained.

Smoking is certainly a hot topic, and you always hit a nerve (just like Hypnoheat among magicians)... All spectators are immediately involved: smokers trying to stop smoking, smokers who successfully stopped, non-smokers bothered by smokers, non-smokers trying to make smokers stop, smokers bothered by non-smokers trying to make them stop etc. Smoking is a real discussion maker, and - to quote Fletcher Knebel - one of the leading causes of statistics. If for some reason this framing does not fit your style or your audience, the gimmicks and suggestive tools can also be applied to a number of alternative presentations (or to effects you may already be performing): simply look at it from a different angle, and you will realize that your imagination is the only limit. The next volume will follow-up accordingly and adopt a less controversial approach, focused on novel gimmicks and politically-correct methods.

For now, thank you again for opening my cellar door... please continue to inspire.

Intoxication is not the wine's fault,
but the man's.

Chinese proverb

A HETEROGENEOUS BIBLIOGRAPHY

(In alphabetical order)

Banachek - *Psychokinetic Silverware* video (2003)

Richard Bandler and John Grinder - *The Structure of Magic* (1975)

Ulf Bolling-Borodin - *Sheherazade* (1999)

Derren Brown - *Pure Effect* (1999)

Jerome Bruner and Cecile Goodman - *Value and Need as Organizing Factors in Perception, in Journal of Abnormal and Social Psychology* (1947)

Richard Busch - *Peek Performances* (2001)

John M. Carroll and Michael K. Tanenhaus - *Prolegomena to a Functional Theory of Word Formation, in Papers from the CLS Parasession on Functionalism* (April 1975)

Tony Corinda - *Thirteen Steps to Mentalism* (1968)

Milton H. Erickson and Ernest L. Rossi - *The Collected Papers of Milton H. Erickson on Hypnosis* (1980)

Dariel Fitzkee - *Magic by Misdirection* (1945)

Marie-Louise von Franz - *Archetypal Dimensions of the Psyche* (1994)

Baldwin R. Hergenhahn and Matthew H. Olson - *An Introduction to Theories of Personality* (1980)

Carl Jung - *Basic Writings* (1959)

Kenton Knepper - *Wonder Words audio series* (1998-2000)

Gary Kurtz - *Leading with your Head* (1992)

André Martinet - *Elements of General Linguistics* (1964)

Giuliana A. L. Mazzoni, Manila Vannucci and Elizabeth F. Loftus - *Misremembering Story Material, in Legal and Criminological Psychology* (February 1999)

Ormond McGill - *The New Encyclopedia of Stage Hypnosis* (1996)

Henning Nelms - *Magic and Showmanship* (1969)

Ivan Pavlov - *The Experimental Psychology and Psychopathology of Animals* (1903)

Judy Perlmutter - *Stop Smoking in Five Days* (1988)

Jon Racherbaumer - *In a Class by Himself: the Legacy of Don Alan* (1996)

George Robinson Jr. - *Hypnoheat/Hot&Cold: the Tin Foil Trick* (1989, rev. 1995)

Tony "Doc" Shiels - *The Cantrip Codex* (1989)

Mark Strivings - *Miracles from the Hip* (2000)

Jon Tremaine - *Close-Up Mental Act* video (2000)

Mark Tyrrell - *How to Use Hypnosis to Help Yourself and Others* audiotape (2002)

INDEX

Underwords provides exclusive material for professional mentalists, hypnotists and psychic entertainers. All publications can be ordered directly from the address below: please contact us for more information, and subscribe to our free newsletter for updates and pre-order discounts on new releases. Thank you for your support!

UNDERWORDS
207 - 1425 Marine Drive
West Vancouver, BC - Canada, V7T 1B9
Tel. (toll-free line): +1-866-308-3388
Fax: +1-604-677-7476
Email: info@underwords.com
Website: http://www.underwords.com

NOTES

Printed in the United Kingdom
by Lightning Source UK Ltd.
123188UK00001B/10-12/A